FROM COSMOS TO CHAOS

FROM COSMOS TO CHAOS

The Science of Unpredictability

Peter Coles

OXFORD
UNIVERSITY PRESS

Great Clarendon Street, Oxford OX2 6DP

Oxford University Press is a department of the University of Oxford.
It furthers the University's objective of excellence in research, scholarship,
and education by publishing worldwide in

Oxford New York

Auckland Cape Town Dar es Salaam Hong Kong Karachi
Kuala Lumpur Madrid Melbourne Mexico City Nairobi
New Delhi Shanghai Taipei Toronto

With offices in

Argentina Austria Brazil Chile Czech Republic France Greece
Guatemala Hungary Italy Japan Poland Portugal Singapore
South Korea Switzerland Thailand Turkey Ukraine Vietnam

Oxford is a registered trade mark of Oxford University Press
in the UK and in certain other countries

Published in the United States
by Oxford University Press Inc., New York

© Oxford University Press 2006

The moral rights of the author have been asserted
Database right Oxford University Press (maker)

First published 2006

All rights reserved. No part of this publication may be reproduced,
stored in a retrieval system, or transmitted, in any form or by any means,
without the prior permission in writing of Oxford University Press,
or as expressly permitted by law, or under terms agreed with the appropriate
reprographics rights organization. Enquiries concerning reproduction
outside the scope of the above should be sent to the Rights Department,
Oxford University Press, at the address above

You must not circulate this book in any other binding or cover
and you must impose the same condition on any acquirer

British Library Cataloguing in Publication Data
Data available

Library of Congress Cataloging in Publication Data

Coles, Peter.
From cosmos to chaos : the science of unpredictability / Peter Coles.
p. cm.
ISBN-13: 978-0-19-856762-2 (alk. paper)
ISBN-10: 0-19-856762-6 (alk. paper)
1. Science—Methodology. 2. Science—Forecasting. 3. Probabilities. I. Title.
Q175.C6155 2006
501′.5192—dc22 2006003279

Typeset by Newgen Imaging Systems (P) Ltd., Chennai, India
Printed in Great Britain
on acid-free paper by
Biddles Ltd, King's Lynn, Norfolk

ISBN 0-19-856762-6 978-0-19-856762-2

1 3 5 7 9 10 8 6 4 2

'The Essence of Cosmology is Statistics'

George Mcvittie

Acknowledgements

I am very grateful to Anthony Garrett for introducing me to Bayesian probability and its deeper ramifications. I also thank him for permission to use material from a paper we wrote together in 1992.

Various astronomers have commented variously on the various ideas contained in this book. I am particularly grateful to Bernard Carr and John Barrow for helping me come to terms with the Anthropic Principle and related matters.

I also wish to thank the publisher for the patience over the ridiculously long time I took to produce the manuscript.

Finally, I wish to thank the Newcastle United defence for helping me understand the true meaning of the word ‘random’.

Contents

List of Figures

Probable Nature

> The true logic of this world is the calculus of probabilities.
>
> James Clerk Maxwell

This is a book about probability and its role in our understanding of the world around us. 'Probability' is used by many people in many different situations, often without much thought being given to what the word actually means. One of the reasons I wanted to write this book was to offer my own perspective on this issue, which may be peculiar because of my own background and prejudices, but which may nevertheless be of interest to a wide variety of people.

My own field of scientific research is cosmology, the study of the Universe as a whole. In recent years this field has been revolutionized by great advances in observational technology that have sparked a 'data explosion'. When I started out as an ignorant young research student 20 years ago there was virtually no relevant data, the field was dominated by theoretical speculation and it was widely regarded as a branch of metaphysics. New surveys of galaxies, such as the Anglo-Australian Two-degree Field Galaxy Redshift Survey (2dFGRS) and the (American) Sloan Digital Sky Survey (SDSS), together with exquisite maps of the cosmic microwave background, have revealed the Universe to us in unprecedented detail. The era of 'precision cosmology' has now arrived, and cosmologists are now realizing that sophisticated statistical methods are needed to understand what these new observations are telling us. Cosmologists have become glorified statisticians.

This was my original motivation for thinking about writing a book, but thinking about it a bit further, I realized that it is not really correct to think that there is anything new about cosmology being a statistic subject. The quote at the start of this book, by the distinguished British mathematician George McVittie actually dates from the 1960s, long before the modern era of rapid data-driven progress.

He was right: cosmology has always been about probability and statistics, even in the days when there was very little data. This is because cosmology is about making inferences about the Universe on the basis of partial or incomplete knowledge; this is the challenge facing statisticians in any context. Looked at in this way, much of science can be seen to be based on some form of statistical or probabilistic reasoning. Moreover, history demonstrates that much of the basic theory of statistics was actually developed by astronomers.

There is also a nice parallel between cosmology and forensic science, which I used as the end piece to my little book *Cosmology: A Very Short Introduction*. We do not do experiments on the Universe; we simply observe it. This is much the same as what happens when forensic scientists investigate the scene of a crime. They have to piece together what happened from the evidence left behind. We do the same thing when we try to learn about the Big Bang by observing the various forms of fallout that it produced. This line of thinking is also reinforced by history: one of the very first forensic scientists was also an astronomer.

These surprising parallels between astronomy and statistical theory are fascinating, but they are just a couple of examples of a very deep connection. It is that connection that is the main point of this book. What I want to explore is why it is so important to understand about probability in order to understand how science works and what it means. By this I mean science in general. Cosmology is a useful vehicle for the argument I will present because so many of the issues hidden in other fields are so obvious when one looks at the Universe as a whole. For example, it is often said that cosmology is different from other sciences because the Universe is unique. Statistical arguments only apply to collections of things, so it is said, so they cannot be applied to cosmology. I do not think this is true. Cosmology is not qualitatively different from any other branch of science. It is just that the difficulties are better hidden in other disciplines.

The attitude of many people towards statistical reasoning is deep suspicion, as can be summarized by the famous words of Benjamin Disraeli: 'There are lies, damned lies, and statistics'. The idea that arguments based on probability are deployed only by disreputable characters, such as politicians and bookmakers, is widespread even among scientists. Perhaps this is partly to do with the origins of the subject in the mathematics of gambling games, not generally regarded as appropriate pastimes for people of good character. The eminent experimental physicist Ernest Rutherford, who split the atom and founded the subject of nuclear physics, simply believed that the use of statistics was a sign of weakness: 'If your experiment needs statistics to analyze the results, you ought to have done a better experiment'.

When I was an undergraduate student studying physics at Cambridge in the early 1980s, my attitude was definitely along the lines of Rutherford's, but perhaps even more extreme. I have never been very good at experiments (or practical things of any kind), so I was drawn to the elegant precision and true-or-false certainty of mathematical physics. Statistics was something practised by sociologists, economists, biologists and the like, not by 'real' scientists. It sounds very arrogant now, but my education both at school and university now definitely promoted the attitude that physicists were intellectually superior to all other scientists. Over the years I have met enough professional physicists to know that this is far from the truth.

Anyway, for whatever reason, I skipped all the lectures on statistics in my course (there were not many anyway), and never gave any thought to the idea I might be missing something important. When I started doing my research degree in theoretical astrophysics at Sussex University, it only took me a couple of weeks to realize that there was an enormous gap in my training. Even if you are working on theoretical matters, if you want to do science you have to compare your calculations with data at some point. If you do not care about testing your theory by observation or experiment then you cannot really call yourself a scientist at all, let alone a physicist. The more I have needed to know about probability, the more I have discovered what a fascinating subject it is.

People often think science is about watertight certainties. As a student I probably thought so too. When I started doing research it gradually dawned on me that if science is about anything at all, it is not about being certain but about dealing rigorously with uncertainty. Science is not so much about knowing the answers to questions, but about the process by which knowledge is increased.

So the central aim of the book is to explain what probability is, and why it plays such an important role in science. Probability is quite a difficult concept for non-mathematicians to grasp, but one that is essential in everyday life as well as scientific research. Casinos and stock markets are both places where you can find individuals who make a living from an understanding of risk. It is strange that the management of a Casino will insist that everything that happens in it is random, whereas the financial institutions of the city are supposed to be carefully regulated. The house never loses, but Stock Market crashes are commonplace.

We all make statements from time to time about how 'unlikely' is for our team to win on Saturday (especially mine, Newcastle United) or how 'probable' it is that it may rain tomorrow. But what do such statements actually mean? Are they simply subjective judgements, or do they have some objective meaning?

In fact the concept of probability appears in many different guises throughout the sciences too. Both fundamental physics and astronomy provide interesting illustrations of the subtle nuances involved in different contexts. The incorporation of probability in quantum mechanics, for example, has led to a widespread acceptance that, at a fundamental level, nature is not deterministic. But we also apply statistical arguments to situations that are deterministic in principle, but in which prediction of the future is too difficult to be performed in practice. Sometimes, we phrase probabilities in terms of frequencies in a collection of similar events, but sometimes we use them to represent the extent to which we believe a given assertion to be true. Also central to the idea of probability is the concept of 'randomness'. But what is a random process? How do we know if a sequence of numbers is random? Is anything in the world actually random? At what point should we stop looking for causes? How do we recognize patterns when there is random noise?

In this book I cut a broad swathe through the physical sciences, including such esoteric topics as thermodynamics, chaos theory, life on other worlds, the Anthropic Principles, and quantum theory. Of course there are many excellent books on each of these topics, but I shall look at them from a different perspective: how they involve, or relate to, the concept of probability. Some of the topics I discuss require a certain amount of expertise to understand them, and some are inherently mathematical. Although I have kept the mathematics to the absolute minimum, I still found I could not explain some concept without using some equations. In most cases I have used mathematical expressions to indicate that something quantitative and rigorous can be said; in such cases algebra and calculus provide the correct language. But if you really cannot come with mathematics at all, I hope I have provided enough verbal explanations to provide qualitative understanding of these quantitative aspects.

So far I have concentrated on the 'official' reasons for writing this book. There is also another reason, which is far less respectable. The fact of the matter is that I quite like gambling, and am fascinated by games of chance. To the disapproval of my colleagues I put £1 on the National Lottery every week. Not because I expect to win but because I reckon £1 is a reasonable price to pay for the little *frisson* that results when the balls are drawn from the machine every Saturday night. I also bet on sporting events, but using a strategy I discovered in the biography of the great British comic genius, Peter Cook. He was an enthusiastic supporter of Tottenham Hotspur, but whenever they played he bet on the opposing team to win. His logic was that, if his own team won he was happy anyway, but if it lost he would receive financial compensation.

As I was writing this book, during the summer of 2005, cricket fans were treated to a serious of exciting contests between England and Australia for one of the world's oldest sporting trophies, The Ashes. The series involved five matches, each lasting five days. After four close-fought games, England led by two games to one (with one game drawn), needing only to draw the last match to win back The Ashes they last held almost 20 years ago. At the end of the fourth day of the final match, at the Oval, everything hung in the balance. I was paralysed by nervous tension. Only a game that lasts five days can take such a hold of your emotions, in much the same way that a five-act opera is bound to be more profound than a pop record. If you do not like cricket you will not understand this at all, but I was in such a state before the final day of the Oval test that I could not sleep. England could not really lose the match, could they? I got up in the middle of the night and went on the Internet to put a bet on Australia to win at 7-1. If England were to lose, I would need a lot of consolation so I put £150 on. A thousand pounds of compensation would be adequate.

As the next day unfolded the odds offered by the bookmakers fluctuated as first England, then Australia took the advantage. At lunchtime, an Australian victory was on the cards. At this point I started to think I was a thousand pounds richer, so my worry about an England defeat evaporated. After lunch the England batsmen came out with renewed vigour and eventually the match was saved. It ended in a draw and England won the Ashes. I had also learned something about myself, that is, precisely how easily I can be bought.

The moral of this story is that if you are looking for a book that tells you how to get rich by gambling, then I am probably not the right person to write it. I never play any game against the house, and never bet more than I can afford to lose. Those are the only two tips I can offer, but at least they are good ones. Gambling does however provide an interesting way of illustrating how to use logic in the presence of uncertainty and unpredictability. I have therefore used this as an excuse for introducing some examples from card games and the like.

2 The Logic of Uncertainty

The theory of probabilities . . . is only common sense reduced to calculus.

Pierre Simon, Marquis de Laplace, A Philosophical Essay on Probabilities

First Principles

Since the subject of this book is probability, its meaning and its relevance for science and society, I am going to start in this chapter with a short explanation of how to go about the business of calculating probabilities for some simple examples. I realize that this is not going to be easy. I have from time to time been involved in teaching the laws of probability to high school and university students, and even the most mathematically competent often find it very difficult to get the hang of it. The difficulty stems not from there being lots of complicated rules to learn, but from the fact that there are so few. In the field of probability it is not possible to proceed by memorizing worked solutions to well known (if sometimes complex) problems, which is how many students approach mathematics. The only way forward is to *think*. That is why it is difficult, and also why it is fun.

I will start by dodging the issue of what probability actually *means* and concentrate on how to use it. The controversy surrounding the interpretation of such a common word is the principal subject of Chapter 4, and crops up throughout the later chapters too. What we can say for sure is that a probability is a number that lies between 0 and 1. The two limits are intuitively obvious. An event with zero probability is something that just cannot happen. It must be logically or physically impossible. An event with unit probability is certain. It must happen, and the converse is logically or physically impossible. In between 0 and 1 lies the crux. You have some idea of what it means to say, for example, that the probability of a fair coin landing heads-up is one-half, or that the probability of a fair dice showing a 6 when you roll it is 1/6. Your understanding of these statements (and others like them) is likely to fall in one or other of the following two basic categories. Either the probability represents what will happen if you toss the coin a large number of times, so that it represents some kind of frequency in a long run of repeated trials, or it is some measure of your assessment of the symmetry (or lack of it) in the situation and your subsequent inability to distinguish possible outcomes. A fair dice has six faces; they all look the same, so there is no reason why any one face should have a higher probability of coming up than any other. The probability of a 6 should therefore be the same as any other face. There are six faces, so the required answer must be 1/6. Whichever way you like to think of probability does not really matter for the purposes of this elementary introduction, so just use whichever you feel comfortable with, at least for the time being. The hard sell comes later.

To keep things as simple as possible, I am going to use examples from familiar games of chance. The simplest involving coin-tossing, rolls of a dice, drawing balls from an urn, and standard packs of playing cards. These are the situations for which the mathematical theory of probability was originally developed, so I am really just following history in doing this.

Let us start by defining an *event* to be some outcome of a 'random' experiment. In this context, 'random' means that we do not know how to predict the outcome with certainty. The toss of a coin is governed by Newtonian mechanics, so in principle, we should be able to predict it. However, the coin is usually spun quickly, with no attention given to its initial direction, so that we just accept the outcome will be randomly either head or tails. I have never managed to get a coin to land on its edge, so we will ignore that possibility. In the toss of a coin, there are two possible outcomes of the experiment, so our event may be either of these. Event A might be that 'the coin shows heads'. Event B might be that 'the coin shows tails'. These are the only two possibilities and they are *mutually exclusive* (they cannot happen at the same time). These two events are also *exhaustive*, in that they represent the entire range of possible outcomes of the experiment. We might as well say, therefore, that the event B is the same as 'not A', which we can denote A^*. Our first basic rule of probability is that

$$P(A) + P(A^*) = 1,$$

which basically means that we can be certain that either something (A) happens or it does not (A^*). We can generalize this to the case where we have several mutually exclusive and exhaustive events: A, B, C, and so on. In this case the sum of all probabilities must be 1: however many outcomes are possible, one and only one of them has to happen.

$$P(A) + P(B) + P(C) + \cdots = 1,$$

This is taking us towards the rule for combining probabilities using the operation 'OR'. If two events A and B are mutually exclusive then the probability of *either A or B* is usually written $P(A \cup B)$. This can be obtained by adding the probabilities of the respective events, that is,

$$P(A \cup B) = P(A) + P(B).$$

However, this is not the whole story because not all events are mutually exclusive. The general rule for combining probabilities like this will have to wait a little.

In the coin-tossing example, the event we are interested in is simply one of the outcomes of the experiment ('heads' or 'tails'). In a throw of a dice, a similar type of event A might be that the score is a 6. However, we might instead ask for the probability that the roll of a dice produces an even number. How do we assign a probability for this? The answer is to reduce everything to the elementary outcomes of the experiment which, by reasons of symmetry or ignorance (or both), we can assume to have equal probability. In the roll of a dice, the six individual faces are taken to be equally probable. Each of these must be assigned a probability of 1/6, so the probability of getting a six must also be 1/6. The probability of getting any even number is found by calculating which of the elementary outcomes lead to this composite event and then adding them together. The possible scores are 1, 2, 3, 4, 5, or 6. Of these 2, 4, and 6 are even. The probability of an even number is therefore given by $P(\text{even}) = P(2) + P(4) + P(6) = 1/2$. There is, of course a quicker way to get this answer. Half the possible throws are even, so the probability must be 1/2. You could imagine the faces of the dice were coloured red if odd and black if even. The probability of a black face coming up would be 1/2. There are various tricks like this that can be deployed to calculate complicated probabilities.

In the language of gambling, probabilities are often expressed in terms of odds. If an event has probability p then the odds on it happening are expressed as the ratio $p:(1-p)$, after some appropriate cancellation. If $p=0.5$ then the odds are $1:1$ and we have an even money bet. If the probability is $1:3$ then the odds are $1/3:2/3$, or after cancelling the threes, $2:1$ against. The process of enumerating all the possible elementary outcomes of an experiment can be quite laborious, but it is by far the safest way to calculate odds.

Now let us complicate things a little further with some examples using playing cards. For those of you who did not misspend your youth playing with cards like I did, I should remind you that a standard pack of playing cards has 52 cards. There are 4 suits: clubs (♣), diamonds (♦), hearts (♥) and spades (♠). Clubs and spades are coloured black, while diamonds and hearts are red. Each suit contains thirteen cards, including an Ace (A), the plain numbered cards (2, 3, 4, 5, 6, 7, 8, 9, and 10), and the face cards: Jack (J), Queen (Q), and King (K). In most games the most valuable is the Ace, following by King, Queen, and Jack and then from 10 down to 2.

Suppose we shuffle the cards and deal one. Shuffling is taken to mean that we have lost track of where all the cards are in the pack, and consequently each one is equally likely to be dealt. Clearly the elementary outcomes number 52 in total, each one being a particular card. Each of these has probability 1/52. Let us try some simple examples of calculating combined probabilities.

What is the probability of a red card being dealt? There are a number of ways of doing this, but I will use the brute-force way first. There are 52 cards. The red ones are diamonds or heart suits, each of which has 13 cards. There are therefore 26 red cards, so the probability is 26 lots of 1/52, or one-half. The simplest alternative method is to say there are only two possible colours and each colour applies to the same number of cards. The probability therefore must be 1/2.

What is the probability of dealing a king? There are 4 kings in the pack and 52 cards in total. The probability must be $4/52=1/13$. Alternatively there are four suits with the same type of cards. Since we do not care about the suit, the probability of getting a king is the same as if there were just one suit of 13 cards, one of which is a king. This again gives 1/13 for the answer.

What is the probability that the card is a red jack or a black queen? How many red jacks are there? Only two: J♦ and J♥. How many black queens are there? Two: Q♣ and Q♠. The required probability is therefore 4/52, or 1/13 again.

What is the probability that the card we pull out is either a red card or a seven? This is more difficult than the previous examples, because it requires us to build a more complicated combination of outcomes. How many sevens are there? There are four, one of each suit. How many red cards are there? Well, half the cards are red so the answer to that question is 26. But two of the sevens are themselves red so these two events are not mutually exclusive. What do we do?

This brings us to the general rules for combining probabilities whether or not we have exclusivity. The general rule for combining with 'or' is

$$P(A \cup B) = P(A) + P(B) - P(A \cap B)$$

The extra bit that has appeared compared to the previous version, $P(A \cap B)$, is the probability of A *and* B both being the case. This formula is illustrated in the figure using a Venn diagram. If you just add the probabilities of events A and B then the intersection (if it exists) is counted twice. It must be subtracted off to get the right answer, hence the result I quoted above.

To see how this formula works in practice, let us calculate the separate components separately in the example I just discussed. First we can directly work out the left-hand side by enumerating the required probabilities. Each card is mutually exclusive of any other, so we can do this straightforwardly. Which cards satisfy the requirement of redness or seven-ness? Well, there are four sevens for a start. There are then two entire suits of red cards, numbering 26 altogether. But two of these 26 are red sevens (7♦ and 7♥) and I have already counted those. Writing all the possible cards down and crossing out the two duplicates leaves 28: two red suits plus two black sevens. The answer for the probability is therefore 28/52 which is 7/13.

Now let us look at the right-hand side. Let A be the event that the card is a seven and B be the event that it is a red card. There are four sevens, so $P(A) = 4/52 = 1/13$. There are 26 red cards, so $P(B) = 26/52 = 1/2$. What we need to know is $P(A \cap B)$, in other words how many of the 52 cards are both red and sevens? The answer is 2, the 7♦ and 7♠, so this probability is $2/52 = 1/26$. The right-hand side therefore becomes $1/13 + 1/2 - 1/26$, which is the same answer as before.

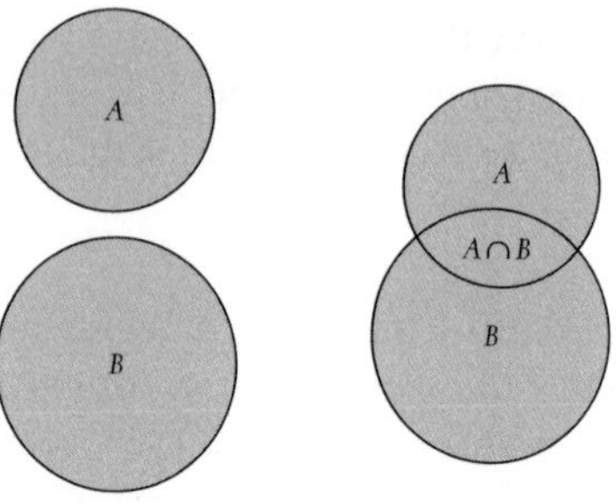

Figure 1 Venn diagrams and probabilities. On the left the two sets A and B are disjoint, so the probability of their intersection is zero. The probability of A or B, P(A∪B) is then just P(A)+P(B). On the right the two sets do intersect so P(A∪B) is given by P(A)+P(B)-P(A∩B).

There is a general formula for the construction of the 'and' probability $P(A \cap B)$, which together with the 'or' formula, is basically all there is to probability theory. The required form is

$$P(A \cap B) = P(A)P(B|A).$$

This tells us the joint probability of two events A and B in terms of the probability of one of them $P(A)$ multiplied by the conditional probability of the second event given the first, $P(B \mid A)$. Conditioning probabilities are probably the most difficult bit of this whole story, and in my experience they are where most people go wrong when trying to do calculations. Forgive me if I labour this point in the following.

The first thing to say about the conditional probability $P(B \mid A)$ is that it need not be the same as $P(B)$. Think of the entire set of possible outcomes of an experiment. In general, only some of these outcomes may be consistent with the event A. If you condition on the event A having taken place then the space of possible outcomes consequently shrinks, and the probability of B in this reduced space may not be the same as it was before the event A was imposed. To see this, let us go back to our example of the red cards and the sevens. Assume that we have picked a red card. The space of possibilities has now shrunk to 26 of the original 52 outcomes. The probability that we have a seven is now just 2 out of 26, or 1/13. In this case $P(A) = 1/2$ for getting a red card, times 1/13 for the conditional probability of getting a seven given that we have a red card. This yields the result we had before.

The second important thing to note is that conditional probabilities are not always altered by the condition applied. In other words, sometimes the event A makes no difference at all to the probability that B will happen. In such cases

$$P(A \cap B) = P(A)P(B).$$

This is a form of the 'and' combination of probabilities with which many people are familiar. It is, however, only a special case. Events A and B are such that $P(B \mid A) = P(B)$ are termed *independent* events. For example, suppose we roll a dice several times. The score on each roll should not influence what happens in subsequent throws. If A is the event that I get a 6 on the first roll and B is that I get a 6 on the second, then $P(B)$ is not affected by whether or not A happens, These events are independent. I will discuss some further examples of such events later, but remember for now that independence is a special property and cannot always be assumed.

The final comment I want to make about conditional probabilities is that it does not matter which way round I take the two events A and B. In other words, 'A and B' must be the same as 'B and A'. This means that

$$P(A \cap B) = P(A)P(B \mid A) = P(B)P(A \mid B) = P(B \cap A).$$

If we swap the order of my previous logic then we take first the event that my card is a seven. Here $P(B) = 1/13$. Conditioning on this event shrinks the space to only four cards, and the probability of getting a red card in this conditioned space is just $P(A \mid B) = 1/2$. Same answer, different order.

A very nice example of the importance of conditional probability is one that did the rounds in university staff common rooms a few years ago, and recently re-surfaced in Mark Haddon's marvellous novel, *The Curious Incident of the Dog in the Night-Time*. In the version with which I am most familiar it revolves around a very simple game show. The contestant is faced with three doors, behind one of which is a prize. The other two have nothing behind them. The contestant is asked to pick a door and stand in front of it. Then the cheesy host is forced to open one of the other two doors, which has nothing behind it. The contestant is offered the choice of staying where he is or switching to the one remaining door (not the one he first picked, nor the one the host opened). Whichever door he then chooses is opened to reveal the prize (or lack of it). The question is, when offered the choice, should the contestant stay where he is, swap to the other door, or does it not matter?

The vast majority of people I have given this puzzle to answer very quickly that it cannot possibly matter whether you swap or not. But it does. We can see why using conditional probabilities. At the outset you pick a door at random. Given no other information you must have a one-third probability of winning. If you choose not to switch, the probability must still be one-third. That part is easy. Now consider what happens if you happen to pick the wrong door first time. That happens with a probability of two-thirds. Now the host has to show you an empty box, but you are standing in front of one of them so he has to show you the other one. Assuming you picked incorrectly first time, the host has been forced to show you where the prize is: behind the one remaining door. If you switch to this door you will claim the prize, and the only assumption behind this is that you picked incorrectly first time around. This means that your probability of winning using the switch strategy is two-thirds, precisely doubling your chances of winning compared with if you had not switched.

Before we get onto some more concrete applications I need to do one more bit of formalism leading to the most important result in this book, *Bayes' theorem*. In its simplest form, for only two events, this is just a rearrangement of the previous equation

$$P(B|A) = \frac{P(B)P(A|B)}{P(A)}.$$

The interpretation of this innocuous formula is the seed of a great deal of controversy about the rule of probability in science and philosophy, but I will refrain from diving into the murky waters just yet. For the time being it is enough to note that this is a theorem, so in itself it is not the slightest bit controversial. It is what you do with it that gets some people upset.

This allows you to 'invert' conditional probabilities, going from the conditional probability of A given B to that of B given A. Here is a simple example. Suppose I have two urns, which are indistinguishable from the outside. In one urn (which with a leap of imagination I will call Urn 1) there are 1000 balls, 999 of which are black and one of which is white. In Urn 2 there are 999 white balls and one black one. I pick an urn and am told it is Urn 1. I prepare to draw a ball from it.

I can assign some probabilities, conditional on this knowledge about which urn it is.

Clearly $P(\text{a white ball} \mid \text{Urn 1}) = 1/1000 = 0.001$ and $P(\text{a black ball} \mid \text{Urn 1}) = 999/1000 = 0.999$. If I had picked Urn 2, I would instead assign $P(\text{a white ball} \mid \text{Urn 2}) = 0.999$ and $P(\text{a black ball} \mid \text{Urn 2}) = 0.001$. So far, so good.

Now I am blindfolded and the urns are shuffled about so I no longer know which is which. I dip my hand into one of the urns and pull out a black ball. What can I say about which urn I have drawn from?

Before going on, I have to suppose that some of you will say that I cannot infer anything. I have discussed this problem many times with students and some just seem to be inextricably welded to the idea that you have to have a large number of repeated observations before you can assign a probability. That is not the case. A draw of one ball is enough to say something in this example. Is not it more likely that the ball came from Urn 1 if it is black?

To do this properly using Bayes' theorem is quite easy. What I want is $P(\text{Urn 1} \mid \text{a black ball})$. I have the conditional probabilities the other way round, so it is straightforward to invert them. Let B be the event that I have drawn from Urn 1 and A be the event that the ball is a black one. I want $P(B \mid A)$ and Bayes' theorem gives this as $P(B)P(A \mid B)/P(A)$. I have $P(A \mid B) = 0.999$ from the previous reasoning. Now I need $P(B)$, the probability that I draw a black ball regardless of which urn I picked. The simplest way of doing this is to say that the urns no longer matter: there are just 2000 balls, 1000 of which are white and 1000 of which are black and they are all equally likely to be picked. The probability is therefore $1000/2000 = 1/2$. Likewise for $P(A)$ the balls do not matter and it is just a question of which of two identical urns I pick. This must also be one-half. The required $P(B \mid A) = 0.999$. If I drew a black ball it is overwhelmingly likely that it came from Urn 1.

This gives me an opportunity to illustrate another operation one can do with probabilities: it is called marginalization. Suppose two events, A and B, like before. Clearly B either does or does not happen. This means that when A happens it is either along with B or not along with B. In other words A must be accompanied either by B or by B^*. Accordingly,

$$P(A) = P(A \cap B) + P(A \cap B^*).$$

This can be generalized to any number of mutually exclusive and exhaustive events, but this simplest case makes the point. The first bit, $P(A \cap B) = P(B)P(A \mid B)$, is what appears on the top of the right-hand side of Bayes' theorem, while the second part is just the probability of getting a black ball given that it is not Urn 1. Assuming nobody sneaked any extra urns in while I was not looking this must be Urn 2. The required inverse probability is then $0.999/(0.999 + 0.001)$, as before.

A common situation where conditional probabilities are important is when there is a series of events that are not independent. Card games are rich sources of such examples, but they usually do not involve replacing the cards in the pack and shuffling after each one is dealt. Each card, once dealt, is no longer available for subsequent deals. The space of possibilities shrinks each time a card is removed from the deck, hence the probabilities shift. This brings us to the difficult business of keeping track of the possibility space for hands of cards in games like poker or bridge. This space can be very large, and the calculations are consequently quite difficult.

In the next chapter I discuss how astronomers and physicists were largely responsible for establishing the laws of probability, but I cannot resist the temptation to illustrate the difficulty of combining probabilities by including here an example which is extremely simple, but which defeated the great French mathematician D'Alembert. His question was: in two tosses of a single coin, what is the probability that heads will appear at least once? To do this problem correctly we need to write down the space of possibilities correctly. If we write heads as H and tails as T then there are actually four possible outcomes in the experiment. In order these are HH, HT, TH, and TT. Each of these has the same probability of one-quarter, which one can reckon by saying that each of these pairs must be equally likely if the coin is fair; there are four of them so the probability must be 1/4. Alternatively the probability of H or T is separately 1/2 so each combination has probability 1/2 times 1/2 or 1/4. Three of the outcomes have at least one head (HH, HT, and TH) so the probability we need is just 3/4. This example is very easy because the probabilities in this case are independent, but D'Alembert still managed to mess it up. When he tackled the problem in 1754 he argued that there are in fact only three cases: heads on the first throw, heads on the second throw, or no heads at all. He took these three cases to be equally likely, and deduced the answer to be 2/3. But they are not equally likely: his first case includes two of the correct cases. His possibilities are mutually exclusive, but they are not equally likely.

As an interesting corollary of D'Alembert's error, consider the following problem. A coin is thrown repeatedly in a sequence. Each result is written down. What is the probability that the pair HT appears in the sequence before TT appears? One's immediate reaction to this is to say, like I did before, that HT and TT must be equally likely, so the probability that the one comes before the other must be just 50%. But this is also wrong, because we are not tossing the coin discrete pairs. It is a continuous sequence in which the pairs overlap and are therefore not independent. Suppose my first throw is a head. That has a probability of 50%. Given this starting point, I have to throw the sequence HT before I get TT. If my first throw is a tail then there are two subsequent possibilities: a head next or a tail next. If I through a head next, I have the sequence TH. Again I have to throw a tail to make TT possible down the line somewhere and that inevitably means I have to have THT before I can get, say, THTT. Only if I throw TT right at the start can I ever get TT before HT. The odds are 3 : 1 against this happening.

Now let us get to the serious business of card games, and what they tell us about permutations and combinations. I will start with Poker, because it is the simplest and probably most popular game to lose money on. Imagine I start with a well-shuffled pack of 52 cards. In a game of five-card draw poker, the players bet on who has the best hand made from five cards drawn from the pack. In more complicated versions of poker, such as Texas hold'em, one has, say, two 'private' cards in one's hand and, say, five on the table in plain view. These community cards are usually revealed in stages, allowing a round of betting at each stage. One has to make the best hand one can using five cards from one's private cards and those on the table. The existence of community cards makes this very interesting because it gives some additional information about other player's holdings. For the present discussion, however, I will just stick to individual hands and their probabilities.

How many possible five-card poker hands are there? To answer this question we need to know about permutations and combinations. Imagine constructing a poker hand from a standard deck. The deck is full when you start, which gives you 52 choices for the first card of your hand. Once that is taken you have 51 choices for the second, and so on down to 48 choices for the last card. One might think the answer is therefore $52 \times 51 \times 50 \times 49 \times 48 = 311875200$, but that is not quite the right answer. It does not actually matter in which order your five cards are dealt to you. Suppose you have four aces and the 2 of clubs in your hand. For example, the sequences (A♠, A♦, A♥, A♣, 2♣) and (A♥, A♣, 2♣, A♥, A♦) are counted separately among the number I obtained above. There is quite a large number of ways of rearranging these five cards amongst themselves whilst keeping the same poker hand. In fact, there are $5 \times 4 \times 3 \times 2 \times 1 = 120$ such permutations. Mathematically this kind of thing is denoted 5!, or five-factorial. Dividing the number above by this gives the actual number of possible poker hands: 2,598,960. This number is important because it describes the size of the 'possibility space'. Each of these hands is an elementary outcome of a poker deal, and each is equally likely.

This calculation is an example of a mathematical combination. The number of combinations one can make of r things chosen from a set of n is usually denoted $C_{n,r}$. In the example above, $r = 5$ and $n = 52$. Note that $52 \times 51 \times 50 \times 49 \times 48$ can be written 52!/47! The general result can be written

$$C_{n,r} = \frac{n!}{r!(n-r)!}.$$

Poker hands are characterized by the occurrence of particular events of varying degrees of probability. For example, a 'flush' is five cards of the same suit but not in sequence (2♠, 4♠, 7♠, 9♠, Q♠). A numerical sequence of cards regardless of suit (e.g. 7♥, 8♦, 9♣, 10♥, J♠) is called a straight. A sequence of cards of the same suit is called a straight flush. One can also have a pair of cards of the same value, three of a kind, four of a kind, or a 'full house' which is three of one kind and two of another.

One can also have nothing at all, that is, not even a pair. The relative value of the different hands is determined by how probable they are.

Consider the probability of getting, say, five spades. To do this we have to calculate the number of distinct hands that have this composition. There are 13 spades in the deck to start with, so there are $13 \times 12 \times 11 \times 10 \times 9$ permutations of five spades drawn from the pack, but, because of the possible internal rearrangements, we have to divide again by 5! The result is that there are 1287 possible hands containing five spades. Not all of these are mere flushes, however. Some of them will include sequences too, for example, 8♠, 9♠, 10♠, J♠, Q♠, which makes them straight flush hands. There are only 10 possible straight flushes in spades (starting with 2, 3, 4, 5, 6, 7, 8, 9, 10 or J). So 1277 of the possible hands are flushes. This logic can apply to any of the suits, so in all there are $1277 \times 4 = 5108$ flush hands and $10 \times 4 = 40$ straight flush hands.

I would not go through the details of calculating the probability of the other types of hand, but I have included a table showing their probabilities obtained by dividing the relevant number of possibilities by the total number of hands at the bottom of the middle column. I hope you will be able to reproduce my calculations!

Type of Hand	**Number of Possible Hands**	**Probability**
Straight Flush	40	0.000015
Four of a Kind	624	0.000240
Full House	3744	0.001441
Flush	5108	0.001965
Straight	10,200	0.003925
Three of a Kind	54,912	0.021129
Two Pair	123,552	0.047539
One Pair	1,098,240	0.422569
Nothing	1,302,540	0.501177
Totals	2,598,960	1.00000

Poker involves rounds of betting in which each player tries to assess how likely his hand is to be at the others involved in the game. If your hand is weak, you can fold and allow the accumulated bets to be given to your opponent. Alternatively, you can bluff.

If you bet heavily on your hand, the opponent may well think it is strong even if it contains nothing, and fold even if his hand has a higher value. To bluff successfully requires a good sense of timing—it depends crucially on who gets to bet first—and extremely cool nerves. To spot when an opponent is bluffing requires real psychological insight. These aspects of the game are in many ways more interesting than the basic hand probabilities, and they are difficult to reduce to mathematics.

Another card game that serves as a source for interesting problems in probability is Contract Bridge. This is one of the most difficult card games to play well because it is a game of logic that also involves chance to some degree. Bridge is a game for four people, arranged in two teams of two. The four sit at a table with the two members of each team opposite each other. Traditionally the different positions are called North, South, East, and West, where North and South are partners, as are East and West.

For each hand of Bridge an ordinary pack of cards is shuffled and dealt out by one of the players, the dealer. Let us suppose that the dealer in this case is South. The pack is dealt out one card at a time starting with West (to dealer's left), then North, and so on in a clockwise direction. Each player ends up with 13 cards.

Now comes the first phase of the game, the auction. Each player looks at his cards and makes a bid, which is essentially a coded message that gives information to his partner about how good his hand is. A bid is basically an undertaking to win a certain number of tricks with a certain suit as trumps (or with no trumps). The meaning of tricks and trumps will become clear later. For example, dealer might bid 'one spade' which is a suggestion that perhaps he and his partner could win one more trick than the opposition with spades as the trump suit. This means winning seven tricks, as there are always 13 to be won in a given deal. The next to bid—in this case West—can either pass 'no bid' or bid higher, like an auction. The value of the suits increases in the sequence clubs, diamonds, hearts and spades. So to outbid one spade, West has to bid at least two hearts, say, if hearts is the best suit for him. Next to bid is South's partner, North. If he likes spades as trumps he can raise the original bid. If he likes them a lot he can jump to a much higher contract, such as four spades (4♠). Bidding carries on in a clockwise direction until nobody dares take it higher, Three successive passes will end the auction, and the contract is established. Whichever player opened the bidding in the suit that was chosen for trumps becomes 'declarer'. If we suppose our example ended in 4♠, then it was South that opened the bidding. If West had opened 2♥ and this had passed round the table, West would be declarer.

The scoring system for Bridge encourages teams to go for high contracts rather than low ones, so if one team has the best cards it does not necessarily get an easy ride. It should undertake an ambitious contract rather than stroll through a simple one. In particular there are extra points for making 'game' (a contract of four spades, four hearts, five clubs, five diamonds, or three no trumps). There is a huge bonus available for bidding and making a grand slam (an undertaking to win all thirteen tricks, that is, seven of something) and a smaller but still impressive bonus for a small slam (six of something).

The second phase of the game now starts. The person to the left of declarer plays a card and the player opposite declarer puts all his cards on the table and becomes 'dummy', playing no further part in this particular hand. Dummy's cards are entirely under the control of the declarer. All three players can see them, but only declarer can see his own hand. The card play is then similar to whist. Each trick consists of four cards played in clockwise sequence from whoever leads. Each player, including dummy, must follow the suit led if he has a card of that suit in his hand. If a player does not have a card of that suit he may 'ruff', that is play a trump card, or simply discard something from another suit. One can win a trick in one of two ways. Either one plays a higher card of the same suit, for example, K♥ beats 10♥. Aces are high, by the way. Alternatively one can play a trump. The highest trump played also wins the trick. Note that more than one player may ruff. For instance, East may ruff only to be over-ruffed by South if both have none of the suit led. Of course one may not have any trumps at all, making a ruff impossible. The possibility of winning a trick by a ruff also does not exist if the contract is of the no-trumps variety. Whoever wins a given trick leads to start the next one. This carries on until 13 tricks have been played. Then comes the reckoning of whether the contract has been made. If so, points are awarded to declarer's team. If not, penalty points are awarded to the defenders. Then it is time for another hand, probably another drink, and very possibly an argument about how badly declarer played the hand.

I have gone through the game in some detail in an attempt to make it clear why this is such an interesting game for probabilistic reasoning. During the auction, partial information is given about every player's holding. It is vital to interpret this information correctly if the contract is to be made. The auction can reveal which of the defending team holds important high cards, or whether the trump suit is distributed strangely. Because the cards are played in strict clockwise sequence this matters a lot. On the other hand, even with very firm knowledge about where the important cards lie, one still often has a difficult logical puzzle to solve if all of one's winners are to be made. It can be a very subtle game.

I have only space here for one illustration of this kind of thing, but it is one that is fun to work out. As is true to a lesser extent in poker, one is not really interested in the initial probabilities of the different hands but rather how to update these probabilities using conditional information as it may be revealed through the auction and card play. In poker this updating is done largely by interpreting the bets one's opponents are making.

Let us suppose that I am South, and I have been daring enough to bid a grand slam in spades (7♠). West leads, and North lays down dummy. I look at my hand and dummy, and realize that we have 11 trumps between us, missing only the King and the 2. I have all other suits covered, and enough winners to make the contract provided I can make sure I win all the trump tricks. The King, however, poses a problem. The Ace of Spades will beat the King, but if I just lead the Ace, it may be that one of East or West has both the K♠ and the 2♠. In this case he would simply play the two to my Ace. The King would be an automatic winner then: as the highest remaining trump it must win a trick eventually. The contract is then lost. Of course if the spades are split 1-1 between East and West then the King drops when I lead the Ace, so that works.

But there is a different way to play this situation. Suppose, for example, that A♠ and Q♠ are on the table and I have managed to win the first trick in my hand. If I think the K♠ lies in West's hand, I lead a spade. West has to play a spade. If he has the King, and plays it, I can cover it with the Ace so it does not win. If, however, West plays low I can play Q♠. This will win if I am right about the location of the King. Next time I can lead the A♠ from dummy and the King falls. This play is called a *finesse*. But is this better than playing for the drop? It is all a question of probabilities, and this in turn boils down to the number of possible deals that allow each strategy to work.

To start with, we need the total number of possible bridge hands. This is quite easy: it is the number of combinations of 13 objects taken from 52, that is $C_{52,13}$. This is a truly enormous number: over 600 billion. You have to play a lot of games to expect to be dealt the same hand twice!

What we now have to do is evaluate the probability of each possible arrangement of the missing King and two. Dummy and declarer's hands are known to me. There are 26 remaining cards whose location I do not know. The relevant space of possibilities is now smaller than the original one. I have 26 cards to assign between East and West. There are $C_{26,13}$ ways of assigning West's 13 cards, but once I have done this the remaining 13 must be in East's hand.

Suppose West has the 2 but not the K. Conditional on this assumption, I know one of his cards, but there are 12 others remaining to be assigned. There are therefore $C_{24,12}$ hands with this possible arrangement of the trumps. Obviously the K has to be with East in this case. The opposite situation, with West having the K but not the 2 has the same number of possibilities associated with it. Suppose instead West does not have any trumps. There are $C_{24,13}$ ways of constructing such a hand: 13 cards from the 24 remaining non-trumps. The remaining possibility is that West has both trumps: this can happen in $C_{24,11}$ ways. To turn these counts into probabilities we just divide by the total number of different ways I can construct the hands of East and West, which is $C_{26,13}$.

Spades in West's Hand	**Number of Hands**	**Probability**	**Drop**	**Finesse**
None	$C_{24,13}$	0.24	0	0
K	$C_{24,12}$	0.26	0.26	0.26
2	$C_{24,12}$	0.26	0.26	0
K2	$C_{24,11}$	0.24	0	0.24
Total	$C_{26,13}$	1.00	0.52	0.50

The last two columns show the contributions of each arrangement to the probability of success of either playing for the drop or the finesse. The drop is slightly more likely to work than the finesse in this case. Note, however, that this ignores any information gleaned from the auction, which could be crucial. Note also that the probability of the drop and the probability of the finesse do not add up to one. This is because there are situations where both could work or both could fail.

This calculation does not mean that the finesse is never the right tactic. It sometimes has much higher probability than the drop, and is often strongly motivated by information the auction has revealed.

Calculating the odds precisely, however, gets more complicated the more cards are missing from declarer's holding. For those of you too lazy to compute the probabilities, the book *On Gambling*, by Oswald Jacoby contains tables of the odds for just about any bridge situation you can think of.

Finally on the subject of Bridge, I wanted to mention a fact that many people think is paradoxical but which is really just a more complicated version of the 'three-door' problem I discussed above. Looking at the table shows that the odds of a 1-1 split in spades here are 0.52 : 0.48 or 13 : 12. This comes from how many cards are in East and West's hands when the play is attempted. There is a much quicker way of getting this answer than the brute force method I used above. Consider the hand with the spade 2 in it. There are 12 remaining opportunities in that hand that the spade K might fill, but there are 13 available slots for it in the other. The odds on a 1-1 split must therefore be 13 : 12. Now suppose instead of going straight for the trumps, I play off a few winners in the side suits (risking that they might be ruffed, of course). Suppose I lead out three Aces in the three suits other than spades and they all win. Now East and West have only 20 cards between them and by exactly the same reasoning as before, the odds of a 1-1 split have become 10 : 9 instead of 13 : 12. Playing out seemingly irrelevant suits has increased the probability of the drop working. Although I have not touched the spades, my assessment of the probability has changed significantly.

I want to end this Chapter with a brief discussion of some more mathematical (as opposed to arithmetical) aspects of probability. I will do this as painlessly as possible using two well-known examples to illustrate the idea of probability distributions and random variables. This requires mathematics that some readers may be unfamiliar with, but it does make some of the examples I use later in the book a little easier to understand.

In the examples I have discussed so far I have applied the idea of probability to discrete events, like the toss of a coin or a ball drawn from an urn. In many problems in statistical science the event boils down to a measurement of something, that is, the numerical value of some variable or other. It might be the temperature at a weather station, the speed of a gas molecule, or the height of a randomly-selected individual. Whatever it is, let us call it X. What one needs for such situations is a formula that supplies the relative probability of the different values X can take. For a start let us assume that X is *discrete*, that is, that it can only take on specific values. A common example is a variable corresponding to a count (the score on a dice, the number of radioactive decays recorded in a second, and so on). In such cases X is an integer, and the possibility space is $\{0, 1, 2, \ldots\}$. In the case of a dice the set is finite $\{1, 2, 3, 4, 5, 6\}$ while in other examples it can be the entire set of integers going up to infinity.

The probability distribution, $p(x)$, gives the probability assigned to each value of X. If I write $P(X=x)=p(x)$ it probably looks unnecessarily complicated, but this means that 'the probability of the random variable X taking on the particular numerical value x is given by the mathematical function $p(x)$'. In cases like this we use the probability laws in a slightly different form. First, the sum over all probabilities must be unity:

$$\sum_x p(x) = 1,$$

If there is such a distribution we can also define the expectation value of X, $E(X)$ using

$$E(X) = \sum_x xp(x)$$

The expectation value of any function of X, say $f(X)$, can be obtained by replacing x by $f(x)$ in this formula so that, for example:

$$E(X^2) = \sum_x x^2 p(x).$$

A useful measure of the spread of a distribution is the variance, usually expressed as the square of the standard deviation, σ, as in

$$\sigma^2(X) = E(X^2) - [E(X)]^2.$$

To give a trivial example, consider the probability distribution for the score X obtained on a roll of a dice. Each score has the same probability, so $p(x)=1/6$ whatever x is. The formula for the expectation value gives

$$\begin{aligned} E(X) &= 1 \times 1/6 + 2 \times 1/6 + 3 \times 1/6 + 4 \times 1/6 + 5 \times 1/6 \\ &\quad + 6 \times 1/6 \\ &= 21/6 = 3.5 \end{aligned}$$

Incidentally, I have never really understood why this is called the expectation value of X. You cannot expect to throw 3.5 on a dice—it is impossible! However, it is what is more commonly known as the average, or arithmetic mean. We can also see that

$$\begin{aligned} E(X^2) &= 1 \times 1/6 + 2^2 \times 1/6 + 3^2 \times 1/6 + 4^2 \times 1/6 + 5^2 \times 1/6 \\ &\quad + 6^2 \times 1/6 \\ &= 91/6 \end{aligned}$$

This gives the variance as $91/6 - (21/6)^2$, which is 35/12. The standard deviation works out to be about 1.7. This is a useful thing as it gives a rough measure of the spread of the distribution around the mean. As a rule of thumb, most of the probability lies within about two standard deviations either side of the mean.

Let us consider a better example, and one which is important in a very large range of contexts. It is called the binomial distribution. The situation where it is relevant is when we have a sequence of n independent 'trials' each of which has only two possible outcomes ('success' or 'failure') and a constant probability of 'success' p. Trials like this are usually called Bernoulli trials, after Daniel Bernoulli who is discussed in the next chapter. We ask the question: what is the probability of exactly x successes from the possible n? The answer is the binomial distribution:

$$p_n(x) = C_{n,x} p^x (1-p)^{n-x}$$

You can probably see how this arises. The probability of x consecutive successes is p multiplied by itself x times, or p^x. The probability of $(n-x)$ successive failures is $(1-p)^{n-x}$. The last two terms basically therefore tell us the probability that we have exactly x successes (since there must be $n-x$ failures). The combinatorial factor in front takes account of the fact that the ordering of successes and failures does not matter. For small numbers n and x, there is a beautiful way called Pascal's triangle, to construct the combinatorial factors. It is cumbersome to use this for large numbers, but in any case these days one can use a calculator.

The binomial distribution applies, for example, to repeated tosses of a coin, in which case p is taken to be 0.5 for a fair coin. A biased coin might have a different value of p, but as long as the tosses are independent the formula still applies. The binomial distribution also applies to problems involving drawing balls from urns: it works exactly if the balls are replaced in the urn after each draw, but it also applies approximately without replacement, as long as the number of draws is much smaller than the number of balls in the urn. It is a bit tricky to calculate the expectation value of the binomial distribution, but the result is not surprising: $E(X) = np$. If you toss a fair coin 10 times the expectation value for the number of heads is 10 times 0.5, which is 5. No surprise there. After another bit of maths, the variance of the distribution can also be found. It is $np(1 - p)$.

The binomial distribution drives me insane every four years or so, whenever it is used in opinion polls. Polling organisations generally interview around 1000 individuals drawn from the UK electorate. Let us suppose that there are only two political parties: Labour and the rest. Since the sample is small the conditions of the binomial distribution apply fairly well. Suppose the fraction of the electorate voting Labour is 40%, then the expected number of Labour voters in our sample is 400. But the variance is $np(1 - p) = 240$. The standard deviation is the square root of this, and is consequently about 15. This means that the likely range of results is about 3% either side of the mean value. The 'term' 'margin of error' is usually used to describe this sampling uncertainty. What it means is that, even if political opinion in the population at large does not change at all the results of a poll of this size can differ by 3% from sample to sample. Of course this does not stop the media from making stupid statements like 'Labour's lead has fallen by 2%'. If the variation is within the margin of error then there is absolutely no evidence that the proportion p has changed at all. Doh!

So far I have only discussed discrete variables. In the physical sciences one is more likely to be dealing with continuous quantities, that is, those where the variable can take any numerical value. Here we have to use a bit of calculus to get the right description: basically, instead of sums we have to use integrals. For a continuous variable, the probability is not located at specific values but is smeared out over the whole possibility space. We therefore use the term probability density to describe this situation. The probability density $p(x)$ is such that the probability that the random variable X takes a value in the range $(x, x + dx)$ is $p(x)\,dx$. The density $p(x)$ is therefore not a probability itself, but a probability *per unit* x. With this definition we can write

$$\int_x p(x)\, dx = 1.$$

The probability that X lies in a certain range, say $[a, b]$, the area under the curve defined by $p(x)$:

$$P(a \leq x \leq b) = \int_a^b p(x)\, dx.$$

Expectation values are defined in an analogous way to the case of discrete variables, but replacing sums with integrals. For example,

$$E(X) = \int_x xp(x)\, dx.$$

I have really included these definitions for completeness. Do not worry too much if you do not know about differential calculus, as I will not be doing anything difficult along these lines. This formalism does however allow me to introduce what is probably the most important distribution in all probability theory. This is the Gaussian distribution, often called the normal distribution. It plays an important role in a whole range of scientific settings. This distribution is described by two parameters: μ and σ, of which more in a moment. The mathematical form is

$$p(x) = \frac{1}{\sigma\sqrt{2\pi}} \exp\left[-\frac{(x-\mu)^2}{2\sigma^2}\right],$$

but it is only really important to recognize the shape, which is the famous 'Bell Curve' shown in the Figure. The expectation value of X is $E[X] = \mu$ and the variance is σ^2.

So why is the Gaussian distribution so important? The answer is found in a beautiful mathematical result called the *Central Limit Theorem*. This used to be called the 'Law of Frequency of Error', but since it applies to many more useful things than errors I prefer the more modern name, This says, roughly speaking, that if you have a variable, X, which arises from the sum of a large number of

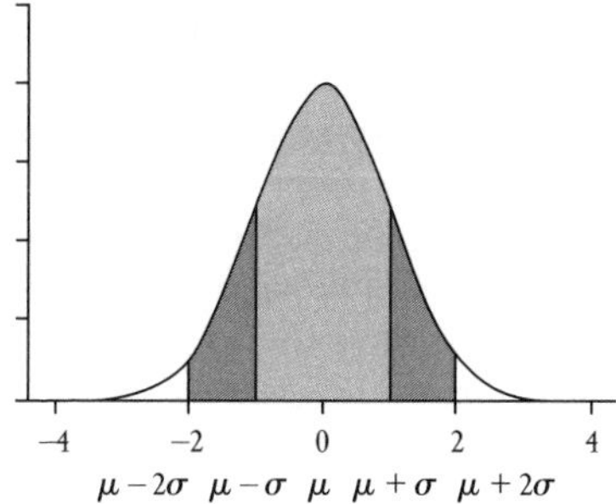

Figure 2 The Normal distribution. The peak of the distribution is at the mean value (μ), with about 95% of the probability within 2σ on either side

independent random influences, so that

$$X = X_1 + X_2 + \cdots X_n$$

then whatever the probabilities of each of the separate influences X_i, the distribution of X will be close to the Gaussian form. All that is required is that the X_i should be independent and there should be a large number of them. Note also that the distribution of the sum of a large number of a independent Gaussian variables is exactly Gaussian. There are an enormous number of situations in the physical and life sciences where some effect is the outcome of a large number of independent causes. Heights of individuals drawn from a population tend to be normally distributed. So do measurement errors in all kinds of experiments. In fact, even the distribution resulting from a very large number of Bernoulli trials tends to this form. In other words, the limiting form of the binomial distribution for a very large n is itself of the Gaussian form, with μ replaced by np and σ^2 replaced by $np(1 - p)$. This does not mean that everything is Gaussian. There are certainly many situations where the central limit theorem does not apply, but the normal distribution is of fundamental importance across all the sciences. The Central Limit Theorem is also one of the most remarkable things in modern mathematics, showing as it does that the less one knows about the individual causes, the surer one can be of some aspects of the result. I cannot put it any better than Sir Francis Galton:

> I know of scarcely anything so apt to impress the imagination as the wonderful form of cosmic order expressed

> by the 'Law of Frequency of Error'. The law would have been personified by the Greeks and deified, if they had known of it. It reigns with serenity and in complete self-effacement, amidst the wildest confusion. The huger the mob, and the greater the apparent anarchy, the more perfect is its sway. It is the supreme law of Unreason. Whenever a large sample of chaotic elements are taken in hand and marshalled in the order of their magnitude, an unsuspected and most beautiful form of regularity proves to have been latent all along.

References and Further Reading

For a good introduction to probability theory, as well as its use in gambling, see:

Haigh, John. (2002). *Taking Chances: Winning with Probability*, Second Edition, Oxford University Press.

A slightly more technical treatment of similar material is:

Packel, Edward. (1981). *The Mathematics of Games and Gambling*, New Mathematical Library (Mathematical Association of America).

More technically mathematical works for the advanced reader include:

Feller, William. (1968). *An Introduction to Probability Theory and Its Applications*, Third Edition, John Wiley & Sons.

Grimmett, G.R. and Stirzaker, D.R. (1992). *Probability and Random Processes*, Oxford University Press.

Jaynes, Ed. (2003). *Probability Theory: The Logic of Science*, Cambridge University Press.

Jeffreys, Sir Harold. (1966). *Theory of Probability*, Third Edition, Oxford University Press.

Simple applications of probability to statistical analysis can be found in

Rowntree, Derek. (1981). *Statistics without Tears*, Pelican Books.

Finally, you must read the funniest book on statistics, once reviewed as 'wildly funny, outrageous, and a splendid piece of blasphemy against the preposterous religion of our time':

Huff, Darrell. (1954). *How to Lie with Statistics*, Penguin Books.

Lies, Damned Lies, and Astronomy

SOCRATES:	Shall we set down astronomy among the subjects of study?
GLAUCON:	I think so, to know something about the seasons, the months and the years is of use for military purposes, as well as for agriculture and for navigation.
SOCRATES:	It amuses me to see how afraid you are, lest the common herd of people should accuse you of recommending useless studies.

Plato, in The Republic

Statistics in Astronomy

Astronomy is about using observational data to test hypotheses about the nature and behaviour of very distant objects, such as stars and galaxies. That immediately sets it apart from experimental disciplines. It is simply impossible to make stars and do experiments with them, even if one could get funding to do it. Nature provides us with a laboratory of a sort, but it also decides what goes on there. We just have to hope that we can observe something that provides us with a way of testing whether our ideas are right. Fortunately, the laboratory we have is enormous and it has a lot going on within it. We observe, measure, catalogue and model (but not necessarily in that order). Eventually patterns emerge, as do rare but decisive exceptions. Models are gradually refined to account for the observations and, hopefully, we end up with some measure of understanding.

As an example of this process, consider how stars work. To the ancients, stars were remote and intangible. The general perception was that they were made of very different material to earthly things and were therefore completely beyond comprehension. Stars are still remote and still intangible, but we now have an almost complete understanding of what they are made of, how they work, and how long they live. We even have a good idea of how stars form, although the details of this process are still far from clear. However, none of this knowledge was gained by taking samples of 'star-stuff', or even following the life-history of individual stars. It takes billions of years for stars to burn their nuclear fuel, and no astronomer has time to watch a star for that long.

The history of stellar evolution theory is long and fascinating, but probably the most important initial breakthrough was the development of a laboratory technique called spectroscopy, pioneered by Robert Wilhelm Bunsen (of burner fame) and Gustav Robert Kirchhoff. This approach involves taking the light emitted by a hot source, such as a flame or an electrical discharge through a gas, and splitting it up using a prism or a diffraction grating. This produces a spectrum showing the familiar pattern of the colours of the rainbow, from red through to blue and violet. White light contains an even blend of all these colours.

What early spectroscopists noticed was that different materials produced light of very specific colours, represented by the appearance of very sharp lines in their spectra. At the time nobody knew why these lines were so sharp—the answer eventually came from quantum theory—but it was realized very early on that the pattern of lines emitted by a particular material in this way was effectively a fingerprint of that material. It was also realized that dark lines could appear in a spectrum, if one were to shine white light through cold matter. These dark lines appear in the same position as the bright lines given off by the same type of material when it is hot. Josef von Fraunhofer had earlier recorded the existence of hundreds of such dark lines when he took a spectrum of the Sun, but it was Kirchhoff who realized that these lines could be identified with the lines produced by the familiar chemical materials he had been playing with in laboratory experiments. Suddenly it became obvious that the Sun was not made of unknown celestial matter, but ordinary stuff. This really was an enormous breakthrough, as it changed the relationship between astronomy and the other sciences forever.

I sometimes get asked to talk to school students thinking about doing an undergraduate degree, and the most common question I get asked on these occasions is 'what is the difference between astronomy and astrophysics?' The distinction is somewhat blurred these days, but there is no doubt that the subject of astrophysics began with

Kirchhoff's realization that stars were made of stuff that could be described by the same laws as terrestrial material. This made it possible to apply the laws of physics to stars and other astronomical objects. Prior to that astronomy had largely consisted of recording the positions and motions of celestial bodies (astrometry), prediction of eclipses and providing navigational tables. But I digress.

The crucial step towards an understanding of stellar evolution was that painstaking observational studies revealed correlations between properties of stars, particularly their temperature and their brightness. There are stars of all different colours, from red to blue. The different colours indicate different temperatures with red being cooler than blue. But stars also differ in brightness from one to another. Independently two astronomers discovered correlations between the colour and brightness of stars. Ejnar Hertzsprung, a Dane working at Potsdam Observatory in Germany first published this correlation but did so in obscure journals. Later on, in 1913, the American Henry Norris Russell of Princeton came to the same conclusion. Their result is encapsulated in one of the most famous diagrams in all science: the Hertzsprung–Russell diagram.

This diagram is usually presented using funny astronomers' units ('magnitude' and 'spectral classification'), but basically it shows brightness up the vertical axis and colour (or temperature) along the horizontal axis. The band lying from top left to bottom right is called the Main Sequence, and its existence was an important spur to theoretical ideas of stellar structure. Our nearest star, the Sun, also lies

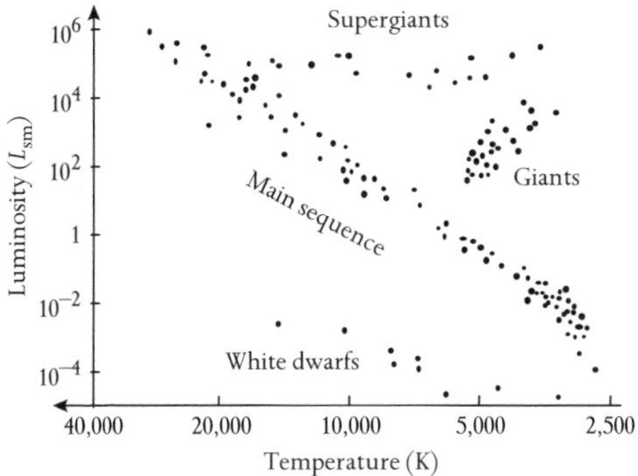

Figure 3 The Hertzsprung-Russell diagram, evidence of a statistical correlation between luminosity and temperature for a population of stars.

on this track. What the diagram shows is that there seems to be a pattern along this sequence: hot stars are bright and faint stars are red. There are two other groups in the diagram, labelled giants (bright and red), and white dwarfs (faint and blue) but these clearly represent different groups. If there were no physical connection between the brightness of a star and its temperature then you would expect the whole diagram to be uniformly sprinkled with points with no discernible pattern or identifiable groups, rather like the following figure:

The fact that the *H-R* diagram does not look like this tells us something important. We now know that all the stars on the main sequence are constructed in the same way. They are enormous balls of gas held together by gravity. In their cores the temperature is so high that they are able to sustain nuclear burning of hydrogen into helium. The heat liberated by these enormous fusion reactors generates pressure that allows them to withstand the force of gravity that tries to make them collapse. They are therefore in a steady equilibrium that can last for billions of years. But not all stars have the same amount of material. Some are larger and more massive than our Sun, some are smaller. It is the mass of a star that determines its position along the Main Sequence. Very big stars are bright and blue, very small ones are faint and red.

The two groups in the H-R diagram labelled 'white dwarfs' and 'giants' are now known to be populated by stars that have finished

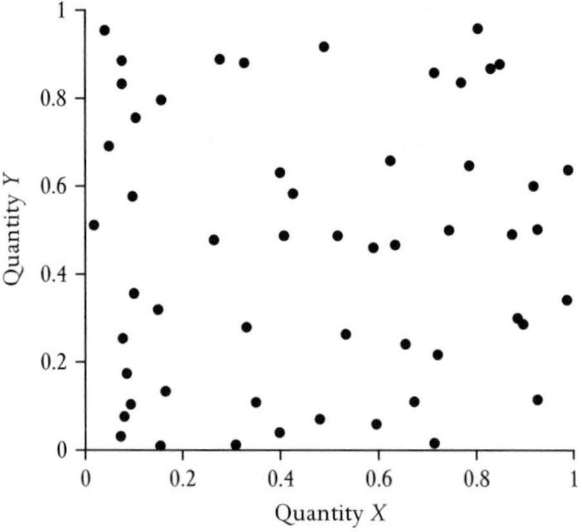

Figure 4 A scatter plot. There is no correlation between the variables X and Y.

their main-sequence lifetime because their cores have run out of hydrogen to burn. Giants are stars where nuclear burning is still going on, but either not in the core or not involving hydrogen. White dwarfs are not burning at all. They are collapsed objects so dense that they are governed by the laws of quantum mechanics. Their heat is not produced continually, but is trapped inside them by their incredible density. It gradually leaks out, so white dwarfs gradually cool and fizzle out.

The existence of the main sequence was a vital clue that enabled theoretical astrophysicists to start constructing physical models of stellar structure. The eventual emergence of the full theory during the twentieth century stands as one of the greatest achievements of natural science. But never forget that it is all based on an elementary application of statistics to relatively simple data.

In these days of powerful telescopes and advanced instrumentation, the observations required to test astronomical theories are often extremely difficult to make, and consequently sometimes prone to sizeable errors and uncertainties. Sometimes these errors are simply *noise*, such as when one is trying to detect a TV signal amongst a buzz of background interference. I have this problem whenever I try to watch anything on Channel 5. Other times the errors are *systematic*, which is a much more difficult state of affairs. An example is when there is an uncertainty in the calibration of a reference source, which can lead to unknown errors in distances to astronomical objects. I will discuss this case in more detail later on. The other important source of error is when there is simply a shortage of data. This was a common occurrence in cosmology only a decade or so ago, but fortunately nowadays the subject is very much data-driven. Whatever the specifics of the observational setup, the ability to make inferences or test hypotheses based on noisy, incomplete or even missing data depends crucially on possession of the correct statistical tools.

This argument has even greater strength when applied to cosmology, which has all the peculiarities of astronomy along with some of its very own thrown in for good measure. In particular, the subject matter of cosmology is the Universe itself, thought of as a single system. Since there is, by definition, only one Universe we have to make inferences about a unique entity. While we may be able to learn about stellar structure by studying populations of stars in various stages of evolution we can not study different universes, at least not observationally. Cosmology therefore takes on some of the character of archaeology or forensic science: we have to learn what we can from what is there in front of us, perhaps a single human bone or scrap of parchment. It may be hard to be certain about inferences made from such relics, but in principle it is both possible and desirable to make them.

Astronomy in Statistics

I hope I have given some credence to the idea that astronomy really does require statistical modes of thinking. However, the communication between astronomy and statistics is by no means a one-way street. The development of the very subject of statistics owes a great deal to astronomers. The history of statistics from the eighteenth century to the present era is a fascinating story recounted in a detailed scholarly book by Hald, so I will restrict myself to edited highlights. To cut a very long story very short, it makes sense to think of the development of the field of statistics as occurring in three distinct revolutionary series, although these are not necessarily in a clean chronological order.

The first stage involved the formulation of the basic laws of probability I described in the previous chapter, almost exclusively inspired by gambling and games of chance. Gambling in various forms has been around for millennia. Sumerian and Assyrian archaeological sites are littered with examples of a certain type of bone, called the astragalus. This is found just above the heel in sheep and deer and its shape means that when it is tossed in the air it can land in any one of four possible orientations. It is therefore the forerunner of modern six-sided dice, and is known to have been used for gambling games as early as 3600 BC. Unlike modern dice, which appeared around 2000 BC, the astragalus is not symmetrical, giving a different probability of it landing in each orientation. It is not thought that there was a mathematical understanding of how to calculate odds in games involving this object or its more symmetrical successors. Games of chance also appear to have been commonplace in the time of Christ—Roman soldiers draw lots at the crucifixion, for example—but there is no evidence of any really formalized understanding of the laws of probability at this time. Playing cards emerged in China sometime during the tenth century BC and were available in western Europe by the fourteenth century. This is an interesting development because playing cards can be used for games involving a great deal of pure skill, as well as an element of randomness. Perhaps it is this aspect that finally got serious intellectuals excited about the probability theory.

The first book on probability that I am aware of was by Cardano. The *Book on Games of Chance* was published in 1663, but written more than a century earlier. Probability theory really got going in 1654 with a famous correspondence between the two famous mathematicians Blaise Pascal and Pierre de Fermat, sparked off by a gambling addict by the name of de Méré. The Chevalier de Méré had played a lot of dice games in his time and felt he had a 'feel' for what was a good bet and what was not. In particular, he had done well financially by betting at even money that he would roll at least one 6 in four rolls of a standard die.

It is quite an easy matter to use the rules of probability to see why he was successful with this game. The odds probability that a single roll of a fair die yields a 6 is 1/6. The probability that it does not yield a 6 is therefore 5/6. The probability that four independent rolls produce no 6s at all is (the probability that the first roll is not a 6) *times* (the probability that the second roll is not a 6) *times* (the probability that the third roll is not a 6) *times* (the probability that the fourth roll is not a 6). Each of the probabilities involved in this multiplication is 5/6, so the result is $(5/6)^4$ which is 625/1296. But this is the probability of losing. The probability of winning is $1 - 625/1296 = 671/1296 = 0.5177$, significantly higher than 50%. It is a good bet.

Unfortunately, so successful had de Méré been that nobody would bet against him any more, and he had to think of another bet to offer. Using his 'feel' for the dice, he reckoned that betting on one or more double-6s in 24 rolls of a pair of dice at even money should also be a winner. Unfortunately for him, he started to lose heavily on this game and in desperation wrote to his friend Pascal to ask why. This set Pascal wondering, and he in turn started a correspondence about it with Fermat. This strange turn of events led not only to the beginnings of a general formulation of probability theory, but also to the binomial distribution and Pascal's Triangle.

The upshot for de Méré was that he abandoned this particular game: the odds on him winning were actually significantly less than fifty per cent. To see this, just consider each throw of a pair of dice as a single 'event'. The probability of getting a double 6 in such an event is 1/36. The probability of not getting a double 6 is therefore 35/36. The probability that a set of 24 independent fair throws of a pair of dice produces no double-6s at all is therefore 35/36 multiplied by itself 24 times, or $(35/36)^{24}$. This is 0.5086 or slightly higher than 50%. The probability that at least one double-six occurs is therefore $1 - 0.5086$, or 0.4914. Our Chevalier has a less than 50% chance of winning, so an even money bet is not a good idea, unless he plans to use this scheme as a tax dodge.

Although both Fermat and Pascal had made important contributions to many diverse aspects of scientific thought as well as pure mathematics, the first real astronomer to contribute to the development of probability in the context of gambling was Christian Huygens, the man who discovered the rings of Saturn in 1655. Two years after his famous astronomical discovery he published a book called *Calculating in Games of Chance*, which introduced the concept of expectation I mentioned in the previous Chapter. Gottfried Wilhelm Leibniz, another famous mathematician who invented the differential calculus independently of (and more elegantly than) Sir Isaac Newton wrote interestingly on the philosophy of probability, especially with respect to its wider ramifications in law, religion and governance. The first phase of the development of statistical theory came to a glorious crescendo with the publication in 1713 of Jakob Bernouilli's wonderful *Ars Conjectandi*.

The first stage of this history already bears the imprint of an astronomer (Huygens), but it is in the second that the deep connection between astronomy and statistics becomes astonishingly clear. Phase two involved the application of probabilistic notions to problems in natural philosophy. Not surprisingly, many of these problems were of astronomical origin but, on the way, the astronomers that tackled them derived some of the basic concepts of statistical theory and practice.

The modern subject of physics began in the sixteenth and seventeenth centuries, although at that time it was usually called Natural Philosophy. The greatest early work in theoretical physics was undoubtedly Newton's great Principia, published in 1687, which presented his idea of universal gravitation which, together with his famous three laws of motion, enabled him to account for the orbits of the planets around the Sun. But majestic though Newton's achievements undoubtedly were, I think it is fair to say that the originator of modern physics was Galileo Galilei.

Galileo was not as much of a mathematical genius as Newton, but he was highly imaginative, versatile and (very much unlike Newton) had an outgoing personality. He was also an able musician, fine artist and talented writer: in other words a true Renaissance man. His fame as a scientist largely depends on discoveries he made with the telescope. In particular, in 1610 he observed the four largest satellites of Jupiter, the phases of Venus and sunspots. He immediately leapt to the conclusion that not everything in the sky could be orbiting the Earth and openly promoted the Copernican view that the Sun was at the centre of the solar system with the planets orbiting around it. The Catholic Church was resistant to these ideas. He was hauled up in front of the Inquisition and placed under house arrest. He died in the year Newton was born (1642).

These aspects of Galileo's life are probably familiar to most readers, but hidden away among scientific manuscripts and notebooks is an important first step towards a systematic method of statistical data analysis. Galileo performed numerous experiments, though he certainly did not carry out the one with which he is most commonly credited. He did establish that the speed at which bodies fall is independent of their weight, not by dropping things off the leaning tower of Pisa but by rolling balls down inclined slopes. In the course of his numerous forays into experimental physics Galileo realized that however careful he was taking measurements, the simplicity of the equipment available to him left him with quite large uncertainties in some of the results. He was able to estimate the accuracy of his measurements using repeated trials and sometimes ended up with a situation in which some measurements had larger estimated errors than others. This is a common occurrence in many kinds of experiment to this day.

Very often the idea is to measure two variables in an experiment, say X and Y. It does not really matter what these two things are, except that X is assumed to be something one can control or measure easily and Y is whatever it is the experiment is supposed to yield information about. In order to establish whether there is a relationship between X and Y one can imagine a series of experiments where X is systematically varied and the resulting Y measured. The pairs of (X, Y) values can then be plotted on a graph like the example shown in Figure 5.

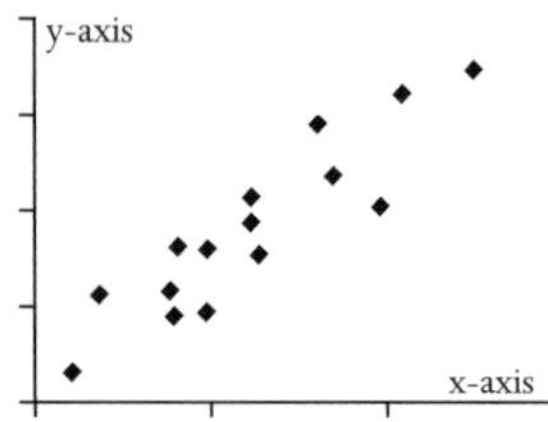

Figure 5 Statistical correlation. Although there is not a perfect mathematical relationship between x and y, there is a tendency for large values of the one to be associated with large values of the other.

In this example on it certainly looks like there is a straight line linking Y and X, but with significant deviations above and below the line caused by the errors in measurement of Y. Thus, you could quite easily take a ruler and draw a line of 'best fit' by eye through these measurements. I spent many a tedious afternoon in the physics labs doing this sort of thing when I was at school. Ideally, though, what one wants is some procedure for fitting a mathematical function to a set of data automatically, without requiring any subjective intervention or artistic skill. Galileo found a way to do this. Imagine you have a set of pairs of measurements (x_i, y_i) to which you would like to fit a straight line of the form $y = mx + c$. One way to do it is to find the line that minimizes some measure of the spread of the measured values around the theoretical line. The way Galileo did this was to work out the sum of the differences between the measured y_i and the predicted values $mx + c$ at the measured values $x = x_i$. He used the absolute difference $|y_i - (mx_i + c)|$ so that the resulting optimal line would, roughly speaking, have as many of the measured points above it as below it. This general idea is now part of the standard practice of data analysis, and as far as I am aware, Galileo was the first scientist to grapple with the problem of dealing properly with experimental error. No doubt Rutherford would just have told him to do a better experiment.

The method used by Galileo was not quite the best way to crack the puzzle, but he had it almost right. It was again an astronomer who provided the missing piece and gave us essentially the same method used by statisticians today. Karl Friedrich Gauss was undoubtedly one of the greatest mathematicians of all time, so it might be objected that he was not really an astronomer. Nevertheless he was director of the Observatory at Göttingen for most of his working life and was a keen observer and experimentalist. In 1809,

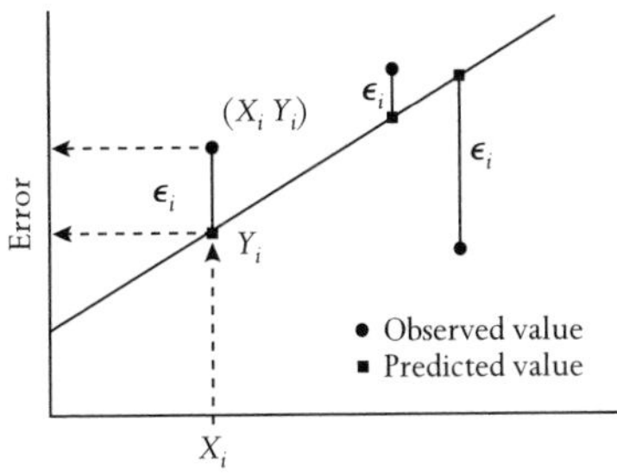

Figure 6 Fitting a line to data. If the data is as shown in Figure 5, no straight line will fit perfectly but one can quantify the mismatch using the residual errors, ε.

he developed Galileo's ideas into the method of least-squares fitting, which is still used today for curve fitting. This is basically the same procedure but involves minimizing the sum of $[y_i - (mx_i + c)]^2$ rather than $|y_i - (mx_i + c)|$. This leads to a much more elegant mathematical treatment of the resulting deviations—the 'residuals'. Gauss also did fundamental work on the mathematical theory of errors in general. The normal distribution I discussed in the previous chapter is often called the Gaussian curve in his honour.

After Galileo, the development of statistics as a means of data analysis in natural philosophy was dominated by astronomers. I can not possibly go systematically through all the significant contributors, but I think it is worth devoting a paragraph or two to a few famous names.

We have already met Jakob Bernoulli, whose famous book on probability was probably written during the 1690s. But Jakob was just one member of an extraordinary Swiss family that produced at least 11 important figures in the history of mathematics. Among them was Daniel Bernoulli who was born in 1700. Along with the other members of his famous family, he had interests that ranged from astronomy to zoology. He is perhaps most famous for his work on fluid flows which forms the basis of much of modern hydrodynamics, especially Bernouilli's principle, which accounts for changes in pressure as a gas or liquid flows along a pipe of varying width.

But the elder Jakob's work on gambling clearly had some effect on Daniel, as in 1735 the younger Bernoulli published an exceptionally clever study involving the application of probability theory to astronomy. It had been known for centuries that the orbits of the planets are confined to the same part in the sky as seen from Earth, a narrow band called the Zodiac. This is because the Earth and the planets orbit in approximately the same plane around the Sun. The Sun's path in the sky as the Earth revolves also follows the Zodiac. We now know that the flattened shape of the Solar System holds clues to the processes by which it formed from a rotating cloud of cosmic debris that formed a disk from which the planets eventually condensed, but this idea was not well established in the time of Daniel Bernouilli. He set himself the challenge of figuring out what the chance was that the planets were orbiting in the same plane simply by chance, rather than because some physical processes confined them to the plane of a protoplanetary disk. His conclusion? The odds against the inclinations of the planetary orbits being aligned by chance were, well, astronomical.

The next 'famous' figure I want to mention is not at all as famous as he should be. John Michell was a Cambridge graduate in divinity who became a village rector near Leeds. His most important idea was the suggestion he made in 1783 that sufficiently massive stars could generate such a strong gravitational pull that light would be unable to escape from them. These objects are now known as black holes (although the name was coined much later by John Archibald Wheeler). In the context of this story, however, he deserves recognition for his use of a statistical argument that the number of close pairs of stars seen in the sky could not arise by chance. He argued that they had to be physically associated, not fortuitous alignments. Michell is therefore credited with the discovery of double stars (or binaries), although compelling observational confirmation had to wait until William Herschel's work of 1803.

I have already mentioned the important role played by Pierre Simon, Marquis de Laplace in the development of statistical theory. His book *A Philosophical Essay on Probabilities*, which began as an introduction to a much longer and more mathematical work, is probably the first time that a complete framework for the calculation and interpretation of probabilities ever appeared in print. First published in 1814, it is astonishingly modern in outlook.

Laplace began his scientific career as an assistant to Antoine Laurent Lavoiser, one of the founding fathers of chemistry. Laplace's most important work was in astronomy, specifically in celestial mechanics, which involves explaining the motions of the heavenly bodies using the mathematical theory of dynamics. In 1796 he proposed the theory

Figure 7 Pierre Simon, Marquis de Laplace, French astronomer and founder of probability theory © Bettman/CORBIS.

that the planets were formed from a rotating disk of gas and dust, which is in accord with the earlier assertion by Daniel Bernouilli that the planetary orbits could not be randomly oriented. In 1776 Laplace had also figured out a way of determining the average inclination of the planetary orbits.

A clutch of astronomers, including Laplace, also played important roles in the establishment of the Gaussian or normal distribution (the 'bell curve' I discussed briefly in the previous chapter). I have also mentioned Gauss's own part in this story, but other famous astronomers played their part. As I mentioned in Chapter 2, the importance of the Gaussian distribution owes a great deal to a mathematical property called the Central Limit Theorem: the distribution of the sum of a large number of independent variables tends to have the Gaussian form. Laplace in 1810 proved a special case of this theorem, and Gauss himself also discussed it at length. A general proof of the Central Limit Theorem was finally furnished in 1838 by another astronomer, Friedrich Wilhelm Bessel, who in the same year was also the first man to measure a star's distance using the method of parallax. Finally, the name 'normal' distribution was coined in 1850 by another astronomer, John Herschel, son of William Herschel.

The final name I wanted to mention here is much more recent and the connection is much less direct, but I wanted to acknowledge Harold Jeffreys who in 1939 published a lovely book on probability that re-kindled an argument about the meaning of probability that had lain dormant for decades. I will discuss the importance of his book in the next chapter. Suffice to say here that Jeffreys was a geophysicist as well as an astronomer and did important work on seismology. He died in 1989.

I hope this makes plain the connections I was trying to establish between statistics and astronomy. And so far, I have only covered two of the three principal stages of statistical history. To complete the story, I should at least explain that the third stage saw the rise of statistical thinking and the life sciences, including sociology and anthropology. I am not at all expert on these areas so I will not discuss them further here, although the wider implications of probabilistic reasoning for society at large will crop up later on in the book. But before getting back to the main thread of my argument, however, I can not resist making one detour down this historical by-way where another surprising encounter awaits.

Forensics

When I give popular talks about cosmology, I sometimes look for appropriate analogies or metaphors in television programmes about forensic science, such as *CSI: Crime Scene Investigation*. I have already mentioned how cosmology is methodologically similar to forensic science because it is generally necessary in both these fields to proceed by observation and inference, rather than experiment and deduction. We cosmologists have only one Universe. Forensic scientists have only one scene of the crime. They can collect trace evidence, look for fingerprints, establish or falsify alibis, and so on. But they can not do what a laboratory physicist or chemist would typically try to do: perform a series of similar experimental crimes under slightly different physical conditions. What we have to do in cosmology is the same as what detectives do when investigating a crime: make inferences and deductions within the framework of a hypothesis that we continually subject to empirical test. This process carries on until reasonable doubt is exhausted, if that ever happens. Of course there is much more pressure on detectives to prove guilt than there is on cosmologists to establish the truth about our Cosmos. That is just as well, because there is still a very great deal we do not know about how the Universe works.

I have a feeling that I have stretched this analogy to breaking point but at least it provides some kind of excuse for mentioning another astronomer-cum-statistician who deserves to be more widely celebrated. Lambert Adolphe Jacques Quételet, a Belgian astronomer who lived from 1796 to 1894. Like Laplace, his principal research interest was in the field of celestial mechanics. He was also an expert in statistics. By the time Quételet was on the scene in the nineteenth century it was not unusual for astronomers to be dab hands at statistics, but he certainly was an outstanding example of this tendency. Indeed, Quételet has been called 'the father of modern statistics'. Amongst other things he was responsible for organizing the first ever international conference on statistics in Paris in 1853. His fame as a statistician owed less to his astronomy, however, than the fact that in 1835 he had written a very influential book called, simply, *On Man*. Quételet had been struck not only by the regular motions performed by the planets across the sky, but also by the existence of strong patterns in social phenomena, such as suicides and crime. If statistics was essential for understanding the former, should it not be deployed in the study of the latter? Quételet's book was an attempt to apply statistical methods to the development of man's physical and intellectual faculties. His follow-up book *Anthropometry, or the Measurement of Different Faculties in Man* (1871) carried these ideas further, at the expense of a much clumsier title. This foray into what he began to call 'social physics' was controversial at the time, for good reason. It also made Quételet extremely famous in his lifetime. The famous statistician Francis Galton wrote about the impact Quételet had on a British lady by the name of Florence Nightingale:

> Her statistics were more than a study, they were indeed her religion. For her Quételet was the hero as scientist, and the presentation copy of his 'Social Physics' is annotated on every page. Florence Nightingale believed—and in all the actions of her life acted on that belief—that the administrator could only be successful if he were guided by statistical knowledge. The legislator—to say nothing of the politician—too often failed for want of this knowledge. Nay, she went further; she held that the universe—including human communities—was evolving in accordance with a divine plan; that it was man's business to endeavour to understand this plan and guide his actions in

> sympathy with it. But to understand God's thoughts, she held we must study statistics, for these are the measure of His purpose. Thus the study of statistics was for her a religious duty.

This type of thinking also spawned a number of highly unsavoury developments in pseudoscience, such as the eugenics movement (in which Galton himself was involved), and some of the vile activities related to it that were carried out in Nazi Germany. But an idea is not responsible for the people who believe in it, and Quételet's work did lead to many good things, such as the beginnings of forensic science. A young medical student by the name of Louis-Adolphe Bertillon was excited by the whole idea of 'social physics', to the extent that he found himself imprisoned for his dangerous ideas during the revolution of 1848, along with one of his professors Achile Guillard, who later invented the subject of demography, the study of racial groups and regional populations. When they were both released, Bertillon became a close confidante of Guillard and eventually married his daughter Zoé. Their second son, Adolphe Bertillon, turned out to be a prodigy, but also something of an *enfant terrible*. Adolphe was so inspired by Quételet's work, no doubt introduced to him by his father, that he hit upon a novel way to solve crimes. He would create a database of measured physical characteristics of convicted criminals. He chose 11 basic measurements, including length and width of head, right ear, forearm, middle and ring fingers, left foot, height, length of trunk, and so on. On their own none of these individual characteristics could be probative, but it ought to be possible to use a large number of different measurements to establish identity with a very high probability. Indeed, after two years' study, Bertillon reckoned that the chances of two individuals having all 11 measurements in common were about four million to one. He added photographs, in portrait and from the side, and a note of any special marks, like scars or moles. *Bertillonage*, as this system became known, was rather cumbersome but proved highly successful in a number of high-profile criminal cases in Paris. By 1892, Bertillon was exceedingly famous but nowadays the word bertillonage only appears in places like the Observer's Azed crossword.

The reason why Bertillon's fame subsided and his system fell into disuse was the development of an alternative and much simpler method of criminal identification: fingerprints. The first systematic use of fingerprints on a large scale was implemented in India in 1858 by a British civil servant attempting to stamp out electoral fraud. It is a pleasing coincidence with which to end this Chapter that his name was William Herschel.

References and Further Reading

This chapter is partly based on an article I wrote for Astronomy and Geophysics in June 2003 ("Statistical Cosmology in Retrospect", A&G, Vol. 44, pp. 16–20).

A comprehensive and scholarly account of the rise of statistical theory is given by

Hald, A. (1988). *A History of Mathematical Statistics from 1750 to 1930*, John Wiley & Sons.

For a superb discussion of the rise of forensic science, see

Wilson, Colin. (1989). *Clues! A History of Forensic Detection*, Warner Books.

Probably the most important essay on probability ever written is

Laplace, Pierre Simon Marquis de, A *Philosophical Essay on Probabilities*, which was first published in 1814. The essay is published in English by Dover (1951).

Bayesians versus Frequentists

> Common language—or, at least, the English language—has an almost universal tendency to disguise epistemological statements by putting them into a grammatical form which suggests to the unwary an ontological statement. A major source of error in current probability theory arises from an unthinking failure to perceive this. To interpret the first kind of statement in the ontological sense is to assert that one's own private thoughts and sensations are realities existing externally in Nature. We call this the 'mind projection fallacy'.
>
> Ed Jaynes, in Probability Theory: The Logic of Science

A Tale of Two Interpretations

So far I have managed to avoid any definitive statement about what probabilities actually mean, concentrating instead on how to manipulate them and how the rules that govern them were gradually uncovered. But now it is time to admit that there is a real controversy surrounding this question. It is not just a debate about the meaning of words: it has real implications for what you think science is about and how scientific reasoning should be pursued.

In a nutshell, there are two competing interpretations of probability. One is the view that is taught most often at an elementary level, that probabilities should be interpreted as frequencies in some large ensemble of repeated experiments under identical conditions. For example, if the probability of a coin turning up heads is 0.5, this means that if I toss such a coin a very large number of times I should find heads showing roughly 50% of the time. Of course in any finite number of trials I will not get an exact 50–50 split between heads and tails. The binomial distribution I discussed in Chapter 2 has a finite spread about its mean value for any value of n, which means that sometimes I will get an above-average number of heads, and sometimes I will be unlucky and get less than the average. Nevertheless, it is true that by making the number n as large as I like, I can make this spread as small as I like and I should get closer and closer in proportion to the expected outcome. Roughly speaking, this goes by the name of the 'law of large numbers' although in the media it is usually called 'the law of averages'. As far as I know there is no such thing as the law of averages.

The rules for probability that I described earlier on in the book apply very neatly if one interprets probability in this way, as proportions in this kind of ensemble. In fact, they are more-or-less trivial when applied to Venn diagrams. The probability of an event A is just the fraction of times A happens in a set of trials. This is also often how one thinks about probability in practice. In a game of Bridge, for example, you might assess the probability of a finesse working to be 50%. This means you might expect to win the trick in about half the hands you play in which the cards you know about have a similar arrangement.

The general term given to this interpretation of probability is *frequentist*. In my experience, it is the interpretation favoured by experimental scientists and observational astronomers. I think this is probably the case because such people are empirically-minded: they really want everything to be observable. By interpreting probability as a frequency of occurrence in repeated experiments it becomes possible, at least in principle, to measure it. One cannot exactly measure a probability, of course, because that would require an infinite ensemble which cannot be constructed in practice. But this is not a very strong objection. No phenomenon can be measured absolutely accurately in any experiment anyway: there is always noise or systematic uncertainty in calibration. The finiteness of any real ensemble merely introduces sampling noise into the problem.

The principal alternative to frequentism is the Bayesian interpretation. To see how this works we need to look at Bayes' theorem, which I will write in a slightly different form to that which I introduced in Chapter 2:

$$P(B|A \cap C) = \frac{P(B|C)P(A|B \cap C)}{P(A|C)}$$

All I have done is to add an extra symbol C representing conditioning information on which all the probabilities now depend. $P(B|C)$ is the probability of B being true given the knowledge of C. The information

C need not be definitely known, but perhaps assumed for the sake of argument. The left-hand side of Bayes' theorem denotes the probability of *B* given both *A* and *C*, and so on. The presence of *C* has not changed anything. This is still a theorem that can be proved to be correct.

Incidentally, although this is called Bayes' theorem the general form of it was actually first written down by Laplace. What Bayes' did was derive the special case of this formula for 'inverting' the binomial distribution. If you remember, this distribution gives the probability of x successes in n independent trials with the same probability p. Bayes was interested in the opposite result: suppose I perform n independent trials and get x successes, what is the probability distribution of p? He got the correct answer, but by very convoluted reasoning. It is quite difficult to justify the name Bayes' theorem, based on what he actually did. This is not the only example in science where the wrong person's name is attached to a result or discovery. In fact, it is almost a law of Nature that any theorem that has a name has the wrong name. I propose that this should henceforth be known as Coles' Law.

Thomas Bayes was born in 1702, son of Joshua Bayes, who was a Fellow of the Royal Society (FRS) and one of the very first non-conformist ministers to be ordained in England. Thomas was himself ordained and for a while worked with his father in the Presbyterian Meeting House in Leather Lane, near Holborn in London. In 1720 he was a minister in Tunbridge Wells, in Kent. He retired from the church in 1752 and died in 1761. Thomas Bayes did not publish a single paper on mathematics during his lifetime but despite this was elected an FRS in 1742. Presumably he had Friends of the Right Sort. The paper containing the theorem that now bears his name was published posthumously in the *Philosophical Transactions of the Royal Society of London* in 1764.

Now comes the controversy. In the frequentist interpretation, *A*, *B*, and *C* are 'events' (e.g. the coin is heads) or 'random variables' (e.g. the score on a dice) attached to which is their probability, indicating their propensity to occur in the imagined ensemble. These things are quite complicated mathematical objects: they do not have specific numerical values, but are represented by a measure over the space of possibilities. They are 'blurred-out' entities. To a Bayesian, the entities *A*, *B*, and *C* have a completely different character. They are logical propositions which can only be either true or false. The entities themselves are not blurred out, but we may have insufficient information to decide which of the two possibilities is correct. In this interpretation, $P(A|C)$ represents the *degree of belief* that it is consistent to hold in the truth of A given the information C. Probability is therefore a generalization of 'normal' deductive logic. In Boolean algebra, the value '0' is associated with a proposition which is false and '1' denotes one that is true. Probability theory is a generalization of this to the intermediate case where there is insufficient information to be certain.

A common objection to Bayesian probability is that it is arbitrary or ill-defined. 'Subjective' is the word that is often bandied about. This is undoubtedly true, at least to the extent that different individuals may have access to different information and therefore assign different probabilities. Given different information C and C' the probabilities $P(A|C)$ and $P(A|C')$ will be different. On the other hand, the same precise rules for assigning and manipulating probabilities apply as before. Identical results should therefore be obtained whether these are applied by any person, or even a robot. The great strength of the Bayesian interpretation is indeed that it all does depend on what information is assumed so this information has to be stated explicitly. The essential assumptions behind a result can be—and, regrettably, often are—hidden in frequentist analyses.

To a Bayesian, probabilities are always conditional on other assumed truths. There is no such thing as an absolute probability, hence my alteration of the form of Bayes's theorem to represent this. A probability such as $P(A)$ has no meaning to a Bayesian: there is always conditioning information. For example, when I blithely assigned a probability of 1/6 to each face on a dice, that assignment was actually conditional on me having no information to discriminate between the appearance of the faces, and no knowledge of the rolling trajectory that would allow me to make a prediction of its eventual resting position.

Probability theory thus becomes not a branch of experimental science but a branch of logic. Like any branch of mathematics it cannot be tested by experiment but only by the requirement that it be internally self-consistent. This brings me to what I think is one of the most important results of twentieth century mathematics, but which is unfortunately almost unknown in the scientific community. In 1946, Cox derived the unique generalization of Boolean algebra under the assumption that such a logic must involve a calculus obtained by associating a single number with any logical proposition. The result he got is beautiful and anyone with any interest in science should make a point of reading his elegant argument. It turns out that the only way to construct a consistent logic of uncertainty incorporating this principle gives exactly the same basic laws of probability I expounded earlier. There is no other way to reason consistently in the face of uncertainty than probability theory. Accordingly, probability theory always applies when there is insufficient knowledge for deductive certainty. Probability is *inductive* logic. This is not just a nice mathematical property. This kind of probability lies at the foundations of a consistent methodological framework that not only encapsulates many common-sense notions about how science works, but also puts at least some aspects of scientific reasoning on a rigorous quantitative footing. This is an important weapon that should be used more often in the battle against the creeping irrationalism one finds in society at large.

I do not want to go into the detailed mathematics of Cox's reasoning, but there is a way of understanding essentially how it works using the so-called *Dutch book* argument. Imagine you are a gambler interested in betting on the outcome of some event. If the game is fair, you would have to expect to pay a stake px to win an amount x if the probability of the winning outcome is p. Now let us imagine that there are several possible outcomes, each with different probabilities, and you are allowed to bet a different amount on each of them. Clearly, the bookmaker has to be careful that there is no combination of bets that guarantees that you will win. Equally, you have to be careful that the bookmaker has not rigged them so that you will always lose.

Now consider a specific example. Suppose there are three possible outcomes; A, B, and C. Your bookie will accept the following bets: a bet on A with a payoff x_A, for which the stake is $p_A x_A$; a bet on B for which the payoff is x_B and the stake $p_B x_B$; and a bet on C with stake $p_C x_C$ and payoff x_C. Think about what happens in the special case where the events A and B are mutually exclusive, and C is just given by $A \cup B$ (A 'or' B). There are then three possible results. First, if A happens but B does not happen, the net return to the gambler is

$$R = x_A(1 - p_A) - x_B p_B + x_c(1 - p_c)$$

where the first term represents the difference between the stake and the return for the successful bet on A, the second is the lost stake corresponding to the failed bet on the event B, and the third term arises from the successful bet on C. Alternatively, if B happens but

A does not happen, the return is

$$R = -x_A p_A + x_B(1 - p_B) + x_c(1 - p_c),$$

constructed in a similar way to the previous result except that the bet on A loses, while those on B and C succeed. Finally there is the possibility that neither A nor B succeeds: in this case the gambler does not win and the return is bound to be negative:

$$R = -x_A p_A - x_B p_B - x_c p_c$$

Notice that A and B cannot both happen because I have assumed that they are mutually exclusive.

Clearly the game is inconsistent if the return is negative whichever of these three outcomes arises. It is a straightforward bit of linear algebra to show that in order to avoid this we need to have

$$det\begin{pmatrix} 1-p_A & -p_B & 1-p_c \\ -p_A & 1-p_B & 1-p_c \\ -p_A & -p_B & -p_c \end{pmatrix} = p_A + p_B - p_c = 0$$

This means that $P(C) = P(A \cup B) = P(A) + P(B)$ which, for the case of two mutually exclusive events A and B yields the sum rule for probabilities. It is the only combination that is consistent from the point of view of betting behaviour. Similar logic leads to the other rules of probability outlined in Chapter 2, including those for events which are not mutually exclusive.

Notice that this kind of consistency has nothing to do with averages over a long series of repeated bets: if the rules are violated then the gambler faces a certain loss in a single outcome of the game.

Sensible betting practice is not always the only criterion when deciding how to deal with uncertain situations. As well as the probability of a particular proposition being true, there are additional factors that may need to be taken into consideration. Any system of logic that assigns more than one number to each proposition will not be equivalent to probability theory. In any case, decision theory is not the same as probability theory. The game of bridge furnishes a nice example. Sometimes one ends up in an over-ambitious contract. In such cases one realizes that the contract can only be made if there is a very unlikely arrangement of the cards held by one's opponents. The choice of cards to play is then generally motivated by the reasoning that 'the only way I can win is if I assume that a very improbable situation is actually the case'. The decision here goes against pure probability, because necessity trumps chance. More generally, human beings often make decisions based not on belief but on desire, and that is something that is very hard to reduce to calculus.

To see how the Bayesian approach works, let us consider a simple example. Suppose we have a hypothesis H (some theoretical idea that we think might explain some experiment or observation). We also have access to some data D, and we also adopt some prior information I (which might be the results of other experiments or simply working assumptions). What we want to know is how strongly the data D supports the hypothesis H given our background assumptions I. To keep it easy, we assume that the choice is between whether H is true or H is false. In the latter case, 'not-H' or H^* is true. If our experiment is at all useful we can construct $P(D|H \cap I)$, the probability that the experiment would produce the data set D if both our hypothesis and the conditional information are true. This is called the *likelihood*, and it involves some knowledge of the statistical errors produced by our measurement. Using Bayes' theorem we can 'invert' this likelihood to give $P(H|D \cap I)$, the probability that our hypothesis is true given the data and our assumptions:

$$P(H|D \cap I) = \frac{P(H|I)P(D|H \cap I)}{P(H|I)P(D|H \cap I) + P(H^*|I)P(D|H^* \cap I)}$$

The right-hand side of this expression is called the posterior probability; the left-hand side involves $P(H|I)$, which is called the prior probability. The principal controversy surrounding Bayesian inductive reasoning involves the prior and how to define it. I will come back to this shortly because it is indeed very important.

This recipe assigns a large posterior probability to a hypothesis for which the product of the prior probability and the likelihood is large. It can be generalized to the case where we want to pick the best of a set of competing hypotheses, say $H_1 \ldots, H_n$. Note that this need not be the set of all possible hypotheses, just those that we have thought about. We can only choose from what is available. The hypotheses may be relatively simple, such as that some particular parameter takes the value x, or they may be composite involving many parameters and/or assumptions. The Big Bang model of our universe, which I discuss in Chapter 7, is a very complicated hypothesis involving at least a dozen parameters which have to be estimated from observations. Anyway, the required result is

$$P(H_i|D \cap I) = \frac{P(H_i|I)P(D|H_i \cap I)}{\sum_j P(H_j|I)P(D|H_j \cap I)}$$

If the hypothesis concerns the value of a parameter—in cosmology this might be the mean density of the Universe—then the allowed space of possibilities is continuous. The sum in the denominator should then be replaced by an integral, but conceptually nothing changes. Our 'best' hypothesis is the one that has the greatest posterior probability. From a frequentist stance the procedure is often instead to maximize the likelihood. According to this approach the best theory is the one that makes the data most probable. This can be the same as the most probable theory, but only if the prior probability is constant.

To give an idea how this works let me go back into frequentist language for a moment. Imagine I have a set of n independent measurements $\{x_i\}$ of a random variable X. Suppose that I know that these are drawn from a Gaussian distribution with mean μ and variance σ^2. These two parameters constitute the model I need to fit to the data. Because the measurements are independent their joint probability can be factored:

$$P(x_1 \cap x_2 \cdots \cap x_n) = P(\{x_i\}) = P(x_1)P(x_2)\cdots P(x_n)$$

The likelihood L is actually $P(\{x_i\} \mid \mu, \sigma)$ but in frequentist fashion I have left the model choice out of this expression, The formula is

$$L(\mu, \sigma) = \frac{1}{(\sigma\sqrt{2\pi})^n} \exp\left(-\frac{1}{2\sigma^2}\sum_i (x_i - \mu)^2\right).$$

The easiest way is to first take the logarithm, which gives

$$-2\log L = \text{const} + 2n \log \sigma + \frac{1}{\sigma^2}\sum_i (x_i - \mu)^2$$

The last term in this expression is usually called χ^2 ('chi-squared'); it is the sum of the squares of the residuals once the expectation value is removed. This explains why Gauss's method of least-squares fitting is such a good idea, at least if the errors are Gaussian. The best estimate for the parameter μ is obtained by differentiating this with respect to the parameter and setting the result to zero. The result is not surprising: it is just the average of the data,

$$\hat{\mu} = \frac{1}{n}\sum_i x_i$$

This is an example of an unbiased estimator: the average of the estimator over an infinite number of repeated trials equal to the parameter value. The maximum-likelihood estimate of the variance is

$$\hat{\sigma}^2 = \frac{1}{n}\sum_i (x_i - \hat{\mu})^2.$$

A Bayesian approach would only produce the same answer if the prior were constant, which it might plausibly well be. However, straightforwardly estimating the mean of a model using the mean of the data, as is implied in this case, can lead to really nonsensical results. Astronomy gives the following nice example. When a supernova explodes it generally produces a burst of very little and difficult-to-detect particles called neutrinos. Most of these pass through the Earth without being recorded, but it is possible to detect a small fraction of them. When the supernova SN1987A exploded (in 1987) a handful of neutrinos from it were indeed detected. As is commonplace in nuclear events, the number of events emerging from the explosion declines from its peak exponentially according to the following law

$$N = N_0 \exp(-t)$$

I have assumed for simplicity that in this the time is measured in seconds, and that the characteristic decay time is 1 s. Suppose we measure arrival times for the three neutrinos at 12, 14, and 16 s timed in a clock in a terrestrial laboratory. The question is: what time on this clock corresponds to the beginning of the explosion? Call this time θ. Obviously no neutrinos can have been emitted before $t=0$, the time of the explosion on the supernova's own clock measured by the radioactive decay law. The time corresponding to this on the laboratory clock simply cannot be any later than 12 s, otherwise atleast one neutrino would have to have made a false start.

The sensible way to approach this problem is first to construct the likelihood; remember that this is the probability of the data given the model. Given a value for θ the probabilities for each arrival time (t_1, t_2, t_3) will be zero for t_i less than θ and have the exponential form given above for t_1 greater than θ. The likelihood is then easy to construct:

$$P(t_1|\theta)P(t_2|\theta)P(t_3|\theta) = \exp[3(\theta - 14)]$$

for $\theta < 12$ and zero everywhere else. The posterior distribution for θ can then be obtained using whatever prior information is available: if this corresponds to a uniform prior then the result is proportional to this likelihood. Notice that the mean arrival time on the lab clock is just $(12 + 14 + 16)/3 = 14$ s.

Unfortunately, the various frequency-inspired approaches to this kind of inference—collectively known as *sampling theory*—can go badly awry. One technique is to make a so-called unbiased estimator, θ^*, which is a combination of the data defined so that its expectation value is equal to θ. The linear unbiased estimator in this particular problem is

$$\theta^* = \frac{1}{3}(t_1 + t_2 + t_3 - 3),$$

but the probability distribution of this given a value of θ has ridiculous properties! It is cut off not at 12, but at the sample mean, 14. It has a maximum at $\theta = 40/3 = 13.333$ and 93.8% of the probability lies in the impossible region above $\theta = 12$.

What has gone wrong here is that we have applied a formula that works for Gaussian (symmetrical) distribution to a case which is very lop-sided. Requiring the estimate to be unbiased in terms of its expectation value really is not the right thing to do. Frequentists will argue that I have used a silly statistic for this particular problem, which is undeniably true. There are better frequentist approaches to this problem which do not give such absurd answers. However, the point I am trying to make is that science is full of examples of people using 'off-the-shelf' sampling-theoretical methods without thinking about the underlying probability theory. It is better to use probability itself than rely on ad hoc approximations to it.

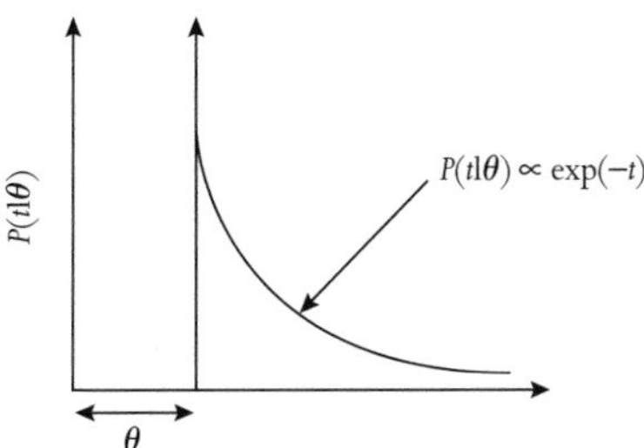

Figure 8 The likelihood for the distribution of arrival times for neutrinos from a supernova explosion for a given value of the event time θ.

This brings us to the difference between 'orthodox' frequency-based statistical methods and the Bayesian alternatives. A standard frequentist approach is to perform what is called a hypothesis test. Here one adopts a hypothesis, usually called the null hypothesis H_0, that concerns some statement we would like to make about a large, perhaps infinite population. The null hypothesis is the model that we would be prepared to accept unless the data tell us otherwise. Then we have some data D. This will in general consist of a set of measurements of some quantity or other obtained from a sample drawn from our population. From the data sample we construct a 'statistic' S which will be some numerical distillation of the set of data D, sometimes in the form of a single number but sometimes a few numbers (such as the sample mean, or the maximum and minimum values found in the sample). Assuming our null hypothesis we can calculate the probability distribution of our statistic: this is the likelihood. Some values of S will be likely in our model, some will be unlikely. What we have to do is choose some 'threshold' probability, usually either 5% or 1%. We reject the null hypothesis if the probability that our measured value of S could arise from a sample drawn from a population whose properties are as assumed within the null hypothesis falls below this value. This critical probability is called the significance level or, more accurately, the 'size' of the test. We can be more confident that we are correctly rejecting H_0 if the chosen threshold is low; a significance level of 5% can also be loosely expressed as a confidence level of 95%.

We might hypothesize, for example, that the mean height of adult males in our population is 1.87 m. We draw a sample and measure its mean to be 1.84 m. Is this consistent with the hypothesis or not? The answer is not a simple 'yes' or 'no': it depends on the probability distribution in the population, the size of the sample, and the significance level chosen.

This type of test may give the wrong answer. We might, for example, reject the null hypothesis when it is actually true. This is called a Type I error, and it will happen with the same probability as our significance level: 5% means that there is a one-in-twenty chance that we goofed. If the population distribution is Gaussian, this level of correspondence corresponds to about 2 standard deviations either side of the mean. It is therefore usually called a '2σ' result. The scientific literature abounds with papers that proudly announce such conclusions, only for them to turn out to be wrong. Indeed, I speculate that many more than 5% of

2σ results are incorrect. This is because this whole idea depends on constructing the sample and making the measurements correctly, which is often harder than it sounds.

The other way we can go wrong with a hypothesis test is to fail to reject the null hypothesis when it is actually false. Unsurprisingly this is called a Type II error. The probability of this happening depends on what the correct hypothesis actually is, so usually we cannot calculate the probability of a Type II error so easily. The word that applies to this aspect of hypothesis testing is called *power*. A powerful test statistic will have a high probability of rejecting a hypothesis H_0 when some alternative is actually true. The power depends on both the null and the alternative hypothesis. Note, however, that there is no implication in any of this that failing to reject the null hypothesis encourages us to believe in it any stronger. The null hypothesis exists only as a straw man.

In a Bayesian approach the logic is quite different. To start with, we always need to specify at least two hypotheses, so there is no preferred 'null'. We must assign their priors and calculate the corresponding likelihoods. Once done we can calculate the posterior probabilities of each hypothesis. Hopefully one will emerge with a higher probability than another, but if it is 50–50 then that is a valid conclusion. The posterior probability emerging from this test can furnish the prior for subsequent ones. If the likelihood for the data given one theory is low, then it is assigned a reduced probability. If the data 'support' a theory then its probability will increase. If the test is not a good one, that is, if the power is low, then the test does not change the relative probabilities of the hypotheses.

Hypothesis testing is just one aspect of orthodox statistical methodology, but it serves to illustrate the difference between it and the Bayesian approach. In the one case, everything is done in 'data space' using likelihoods, and in the other we work throughout with probabilities of hypothesis.

As I mentioned above, it is the presence of the prior probability in the general formula that is the most controversial aspect of the Bayesian approach. The attitude of frequentists is often that this prior information is completely arbitrary or at least 'model-dependent'. Being empirically-minded people, by and large, they prefer to think that measurements can be made and interpreted without reference to theory at all. Moreover, it has to be said that no entirely rigorous and generally applicable method of assigning prior probabilities is known anyway. This is a big problem for Bayesians, but there are many situations where it is known how to assign priors uniquely and consistently, and there is a general principle—called maximum entropy—which may well eventually yield a definitive answer. The reason why 'MaxEnt' is not quite universally accepted among scientists is that it is difficult and sometimes impractical to apply, and sometimes just reveals that the problem one is trying to solve is ill-posed.

In a nutshell, the maximum entropy principle involves the assignment of measure of the lack of information contained in a probability distribution. We will come up against entropy again later on, in Chapter 6, but for the moment we can take it to represent the lack of information carried by the probability distribution. The maximum entropy distribution is the choice of this distribution that carries least information of all. It is therefore the least prejudiced way of assigning a prior given our conditioning information. It adds no further prejudice to what we have adopted at the outset in our choice of information. In the case where there are a discrete number of possibilities each of which has probability p_i, the entropy is

$$S = -\sum_i p_i \log \frac{p_i}{m_i},$$

where m_i is a suitable measure over the possibility space. For the case of a continuous parameter x, for example, it is

$$S = -\int p(x) \log \frac{p(x)}{m(x)}\, dx.$$

The entropy must be maximized subject to the constraint that the probabilities add up to 1; this is called normalization. There may also be other constraints imposed by other bits and pieces of information we have. In the examples about probability I gave in Chapter 2, the appropriate measure is uniform but in complex problems it need not be. Even identifying the appropriate 'hypothesis space' can be difficult. If we know the appropriate measure $m(x)$ and have no other constraints other than normalization, then the MaxEnt principle assigns a constant prior over the hypothesis space. This reproduces what Laplace called 'The Principle of Indifference'. Often, however, we have more to go on than this. Symmetry properties can lead us away from a uniform prior. For example, if we think that the appropriate measure possesses some invariance with respect to scale rather than location then the appropriate measure should be uniform not in x but in $\log x$. In the absence of any other constraints, maximum entropy then gives a probability distribution that is of the form $p(x) \propto 1/x$. This is often called the Jeffrey's prior. In more complex problems, the symmetries present may be very subtle and the MaxEnt procedure subsequently hard to apply. I will discuss an example of this in the context of cosmology later on, in Chapter 8.

Some Bayesian probabilists have given up on the idea that priors can be assigned in a systematic and rigorous fashion. They accept that the choice of prior is entirely subjective. This form of weak Bayesianism is a kind of half-way house, but in the absence of any universally applicable objective method for assigning priors, it is perhaps understandable. Moreover it does at least force one to admit what particular choice of prior one is using, something which is never made explicit in a frequentist analysis. A subjectively-chosen prior may be little more than prejudice, but at least the Bayesian system forces one to put one's prejudices on the table.

As an aside, I should mention an alternative distribution of the Gaussian distribution using maximum entropy. Suppose that in addition to normalization, we constrain our maximization procedure using the additional requirements that:

$$\int xp(x)\,dx = \mu$$

and

$$\int (x-\mu)^2 p(x)\,dx = \sigma^2$$

In other words we specify the mean and variance. If we take the measure $m(x)$ to be uniform then the application of MaxEnt produces the Gaussian distribution obtained in Chapter 2. This is therefore the 'most random' distribution having a fixed mean and variance.

Assuming we can assign the prior probabilities in an appropriate way what emerges from these considerations is a consistent methodology for scientific progress. The scheme starts with the hardest part—theory creation. This requires human intervention, since we have no automatic procedure for dreaming up hypotheses from thin air.

Once we have a set of hypotheses we need data against which theories can be compared using their relative probabilities. The experimental testing of a theory can happen in many stages: the posterior probability obtained after one experiment can be fed in, as prior, into the next. The order of experiments does not matter. This all happens in an endless loop, as models are tested and refined by confrontation with experimental discoveries, and are forced to compete with new theoretical ideas. Often one particular theory emerges as most probable for a while, such as in particle physics where a 'standard model' has been in existence for many years. But this does not make it absolutely right; it is just the best bet amongst the alternatives. Likewise the Big Bang model does not represent the absolute truth, but is just the best available model in the face of the manifold relevant observations we now have concerning the Universe's origin and evolution. The crucial point about this methodology is that it is inherently inductive: all the reasoning is carried out in 'hypothesis space' rather than 'observation space'. Science is, essentially, inverse reasoning.

Another important feature of Bayesian reasoning is that it gives precise motivation to things that we are generally taught as rules of thumb. The most important of these is Ockham's Razor. This famous principle of intellectual economy is variously presented in Latin as *Pluralites non est ponenda sine necessitate* or *Entia non sunt multiplicanda praetor necessitatem.* Either way, it means basically the same thing: the simplest

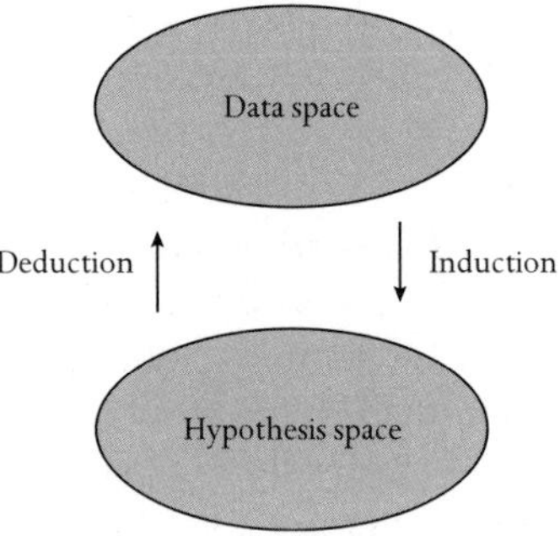

Figure 9 Inductive versus deductive logic. In deductive schemes one uses a theory or model to calculate what should be observed in reality. Testing the model is done in data space. Using inductive logic is different: one uses data to make statements about the probability of the model. Testing in this case is done in hypothesis space.

theory which fits the data should be preferred. William of Ockham, to whom this dictum is attributed, was an English Scholastic philosopher (probably) born at Ockham in Surrey in 1280. He joined the Franciscan order around 1300 and ended up studying theology in Oxford. He seems to have been an outspoken character, and was in fact summoned to Avignon in 1323 to account for his alleged heresies in front of the pope, and was subsequently confined to a monastery from 1324 to 1328, He died in 1349. It is not known how frequently be showed.

In the framework of Bayesian inductive inference, it is possible to give precise reasons for adopting Ockham's razor. In the presence of noise, which is inevitable, there is bound to be some sort of trade-off between goodness-of-fit and simplicity. If there is a lot of noise then a simple model is better: there is no point in trying to reproduce every bump and wiggle in the data with a new parameter or physical law. On the other hand if there is very little noise, every feature in the data is real and your theory fails if it cannot explain it. It is useful to consider what happens when we generalize one theory by adding to it some extra parameters. Suppose we begin with a very simple theory, just involving a parameter x, but we fear it may not fit the data. We therefore add a couple of more parameters, say y and z. We do not know the appropriate numerical values at the outset, so we must infer them by comparison with the available data. Such quantities are usually called 'floating' parameters; there are many in the Big Bang model, for example. Obviously, having three degrees of freedom with which to describe the data should enable one to get a closer fit than is possible with just one. The greater flexibility within the general theory can be exploited to match the measurements more closely than the original. In other words, such a model can improve the likelihood factor. But there is a price to be paid. Each new parameter has to have a prior probability assigned to it. This probability will generally be smeared out over a range of values where the experimental results subsequently show that the parameters do not lie. Even if the extra parameters allow a better fit to the data, this dilution of the prior probability may result in the posterior probability being lower for the generalized theory than the simple one. The more parameters are involved, the bigger the space of prior possibilities for their values, and the harder it is for the improved likelihood to win out. Arbitrarily complicated theories are simply improbable.

The best theory is the most probable one, for which the product of likelihood and prior is largest.

To give a more quantitative illustration of this consider a given model M which has a set of N floating parameters represented as a vector $\lambda = (\lambda_1, \lambda_2, \ldots, \lambda_N)$. In a sense each choice of parameters represents a different model, or member of the family of models labelled M. We have some data D and can consequently form a likelihood function $P(D|\underline{\lambda}, M)$. In Bayesian fashion we have to assign a prior probability to the parameters of the model $P(\underline{\lambda}|M)$ which, if we are being honest, we should do in advance of making any measurements. The interesting thing to look at now is not the best-fitting choice of model parameters but the extent to which the data support the model: this is encoded in a sort of average of likelihood over the prior probability space:

$$P(D|M) = \int P(D|\underline{\lambda}, M) P(\underline{\lambda}|M)\, d^N \underline{\lambda}.$$

This function is sometimes called the 'evidence'. Its usefulness emerges when we ask the question whether our N parameters are sufficient to get a reasonable fit to the data. Should we add another one to improve things a bit further? And why not another one after that? When should we stop? The answer is that although adding an extra degree of freedom can increase the first term in the integral (the likelihood), it also suffers a penalty in the second factor. If the improvement in fit is marginal and/or the data is noisy, then the second factor wins and the evidence for the $(N+1)$-parameter model is smaller than the N-parameter version. Ockham's razor has done its job.

This is a satisfying result that is in nice accord with common sense. But I think it goes much further than that. Many modern-day physicists are obsessed with the idea of a 'Theory of Everything' or TOE. Such a theory would entail the unification of all physical theories—all laws of Nature, if you like—into a single principle. An equally accurate description would then be available, in a single formula, of phenomena that are currently described by distinct theories with separate sets of parameters. Instead of textbooks on mechanics, quantum theory, gravity, electromagnetism, and so on, physics students would need just one book. The physicist Stephen Hawking has described the quest for a TOE as like trying to read the Mind of God. I think that is silly. If a TOE is ever constructed, it will be the most economical available description of the Universe. Not the Mind of God, just the best way we have of saving paper.

So what are the main differences between the Bayesian and frequentist views? First, I think it is fair to say that the Bayesian framework is enormously more general than is allowed by the frequentist notion that probabilities must be regarded as relative frequencies in some ensemble, whether that is real or imaginary. In the latter interpretation, a proposition is at once true in some elements of the ensemble and false in others. It seems to me to be a source of great confusion to substitute a logical AND for what is really a logical OR. The Bayesian stance is also free from problems associated with the failure to incorporate in the analysis any information that cannot be expressed as a frequency. Would you really trust a doctor who said that 75% of the people she saw with your symptoms required an operation, but who did not bother to look at your own medical files?

As I mentioned above, frequentists tend to talk about random variables. To a Bayesian there are no random variables, only variables whose values we do not know. A random process is simply one about which we only have sufficient information to specify probability distributions rather than definite values.

More fundamentally, it is clear from the fact that the combination rules for probabilities were derived by Cox uniquely from the requirement of logical consistency, that any departure from these rules will, generally speaking, involve logical *inconsistency*. Many of the standard statistical data analysis techniques—including the simple 'unbiased estimator' mentioned briefly above—used when the data consist of repeated samples of a variable having a definite but unknown value, are not equivalent to Bayesian reasoning. These methods can, of course, give good answers, but they can all be made to look completely silly by suitable choice of dataset. Ed Jaynes' book *Probability Theory: The Logic of Science* gives numerous examples. Repeated samples comprise proportions, not probabilities.

By contrast, I am not aware of any example of a paradox or contradiction that has ever been found using the correct application of Bayesian methods, although method can be applied incorrectly. Furthermore, in order to deal with unique events like the weather, frequentists are forced to introduce the notion of an ensemble, a perhaps infinite collection of imaginary possibilities, to allow them to retain the notion that probability is a proportion. Provided the calculations are done correctly, the results of these calculations should agree with the Bayesian answers. On the other hand, frequentists often talk about the ensemble as if it were real. This is dangerous. There is only one, imperfectly known, system.

It is ironic that the pioneers of probability theory, principally Laplace, unquestionably adopted a Bayesian rather than frequentist interpretation for their probabilities. Frequentism arose during the nineteenth century and held sway until recently. I recall giving a conference talk about Bayesian reasoning only to be heckled by the audience with comments about 'new-fangled, trendy Bayesian methods'. Nothing could have been less apt. Probability theory predates the rise of sampling theory and all the frequentist-inspired techniques that modern-day statisticians like to employ.

Most disturbing of all is the influence that frequentist and other non-Bayesian views of probability have had upon the development of a philosophy of science, which I believe has a strong element of inverse reasoning or inductivism in it. The argument about whether there is a role for this type of thought in science goes back at least as far as Roger Bacon who lived in the twelfth century. Much later the brilliant Scottish empiricist philosopher and enlightenment figure David Hume argued strongly against induction. Most modern anti-inductivists can be traced back to this source. Pierre Duhem has argued that theory and experiment never meet face-to-face because in reality there are hosts of auxiliary assumptions involved in making this comparison. This is nowadays called the Quine–Duhem thesis. Actually, for a Bayesian this does not pose a logical difficulty at all. All one has to do is set up prior distributions for the required parameters, calculate their posterior probabilities and then integrate over those that are not related to measurements. This is just an expanded version of the idea of marginalization that I introduced in Chapter 2.

Carnap, a logical positivist, attempted to construct a complete theory of inductive reasoning which bears some relationship to Bayesian thought, but he failed to apply Bayes' theorem in the correct way. Carnap distinguished between two types or probabilities—logical and factual. Bayesians do not—and I do not—think this is necessary. The definition I described above seems to me to be quite coherent on its own. Other philosophers of science reject the notion that inductive reasoning has any epistemological value at all. This anti-inductivist stance, often somewhat misleadingly called deductivist (irrationalist would be a better description), is evident in the thinking of three of the most influential philosophers of science of the last century: Karl Popper, Thomas Kuhn and, most recently, Paul Feyerabend. Regardless of the ferocity of their arguments with each other, these have in common that at the core of their systems of thought lies the rejection of all forms of inductive reasoning. The line of thought that ended in this intellectual cul-de-sac began with the brilliant work of the Scottish empiricist philosopher David Hume. For a thorough analysis of the anti-inductivists mentioned above and their obvious debt to Hume, see David Stove's book *Popper and After: Four Modern Irrationalists.* I will just make a few inflammatory remarks here.

Karl Popper really began the modern era of science philosophy with his *Logik der Forschung*, which was published in 1934. There is not really much about probability theory in this work, which is strange for a work which claims to be about the logic of science. Popper also managed to, on the one hand, accept probability theory, but on the other, to reject induction. I find it therefore very hard to make sense of his work at all. It is also clear that, at least outside Britain, Popper is not really taken seriously by many people as a philosopher. Inside Britain it is very different and I'm not at all sure I understand why. In my experience, most working physicists seem to subscribe to some version of Popper's basic philosophy. Among the things Popper has claimed is that all observations are 'theory-laden' and that 'sense-data, untheoretical items of observation, simply do not exist'. I do not think it is possible to defend this view, unless one asserts that numbers do not exist. Data are numbers. They can be incorporated in the form of propositions about parameters in any theoretical framework we like. It is of course true that the possibility space is theory-laden. It is a space of theories, after all. Theory does suggest what kinds of experiment should be done and what data is likely to be useful. But data can be used to update probabilities of anything.

Popper has also insisted that science is deductive rather than inductive. Part of this claim is just a semantic confusion. It is necessary at some point to deduce what the measurable consequences of a theory might be before one does any experiments. He does, however, reject the basic application of inductive reasoning in updating probabilities in the light of measured data. He asserts that no theory ever becomes more probable when evidence is found in its favour. Every scientific theory begins infinitely improbable, and is doomed to remain so.

Now there is a grain of truth in this, or can be if the space of possibilities is infinite. Standard methods for assigning priors often spread the unit total probability over an infinite space, leading to a prior probability which is formally zero. This is the problem of improper priors. But this is not a killer blow to Bayesianism. Even if the prior is not strictly normalizable, the posterior probability can be. In any case, given sufficient relevant data the cycle of experiment-measurement-update of probability assignment usually soon leaves the prior far behind. Data usually count in the end.

The idea by which Popper is best known is the dogma of falsification. According to this doctrine, a hypothesis is only said to be scientific if it is capable of being proved false. In real science certain 'falsehood' and certain 'truth' are almost never achieved. Theories are simply more probable or less probable than the alternatives on the market. The idea that experimental scientists struggle through their entire life simply to prove theorists wrong is a very strange one, although I definitely know some experimentalists who chase theories like lions chase gazelles. To a Bayesian the right criterion is not falsifiability but testability, the ability of the theory to be rendered more or less probable using further data. Nevertheless, scientific theories generally do have untestable components. Any theory has its *interpretation*, which is the untestable baggage that we need to supply to make it comprehensible to us. But as long as it can be tested, it can be scientific.

Popper's work on the philosophical ideas that ultimately led to falsificationism began in Vienna, but the approach subsequently gained enormous popularity in western Europe. The American Thomas Kuhn later took up the anti-inductivist baton in his book *The Structure of Scientific Revolutions*. Kuhn is undoubtedly a first-rate historian of science and this book contains many perceptive analyses of episodes in the development of physics. His view of scientific progress is cyclic. It begins with a mass of confused observations and controversial theories, moves into a quiescent phase when one theory has triumphed over the others, and lapses into chaos again when the further testing exposes anomalies in the favoured theory. Kuhn coined the word *paradigm* to describe the model that rules during the middle stage.

The history of science is littered with examples of this process, which is why so many scientists find Kuhn's account in good accord with their experience. But there is a problem when attempts are made to fuse this historical observation into a philosophy based on anti-inductivism. Kuhn claims that we 'have to relinquish the notion that changes of paradigm carry scientists . . . closer and closer to the truth.' Einstein's theory of relativity provides a closer fit to a wider range of observations than Newtonian mechanics, but in Kuhn's view this success counts for nothing.

Paul Feyerabend has extended this anti-inductivist streak to its logical (though irrational) extreme. His approach has been dubbed 'epistemological anarchism', and it is clear that he believed that all theories are equally wrong. He is on record as stating that normal science is a fairytale, and that equal time and resources should be spent on 'astrology, acupuncture and witchcraft'. He also categorized science alongside 'religion, prostitution, and so on'. His thesis is basically that science is just one of many possible internally consistent views of the world, and that the choice between which of these views to adopt can only be made on socio-political grounds.

Feyerabend's views could only have flourished in a society deeply disillusioned with science. Of course, many bad things have been done in science's name, and many social institutions are deeply flawed. One cannot expect anything operated by people to run perfectly. It's also quite reasonable to argue on ethical grounds which bits of science should be funded and which should not. But the bottom line is that science does have a firm methodological basis which distinguishes it from pseudo-science, the occult and new age silliness. Science is distinguished from other belief-systems by its rigorous application of inductive reasoning and its willingness to subject itself to experimental test. Not all science is done properly, of course, and bad science is as bad as anything.

The Bayesian interpretation of probability leads to a philosophy of science which is essentially epistemological rather than ontological. Probabilities are not 'out there' but in our minds, representing our imperfect knowledge and understanding. Scientific theories are not absolute truths. Our knowledge of reality is never certain. But we are able to reason consistently about which of our theories provides the best available description of what is known at any given time. If that description fails when more data are gathered, we move on, introducing new elements or abandoning the theory for an alternative. This process go on forever: there may never be a final theory. The game might have no end. But at least we know the rules.

References and Further Reading

A must-read book for anyone seriously interested in the Bayesian interpretation of probability, especially those in oppositions, is:

Jaynes, Ed. (2003). *Probability Theory: The Logic of Science*, Cambridge University Press.

The derivation of probability theory as the method of inductive logic is

Cox, R.T. (1946). Probability, Frequency and Reasonable Expectation, *American Journal of Physics*, 14, 1–13.

Various philosophical approaches to probability in the philosophy of science are discussed in:

Carnap, R. (1950). *Logical Foundations of Probability*, University of Chicago Press.

Carnap, R. (1952). *The Continuum of Inductive Methods*, University of Chicago Press.

Popper, K.R. (1959). *The Logic of Scientific Discovery*, Hutchinson.

Kuhn, T.S. (1970). *The Structure of Scientific Revolutions*, Second Edition, University of Chicago Press.

Feyerabend, P.K. (1975). Against *Method: Outline of an Anarchistic Theory of Knowledge*, New Left Books.

Feyerabend, P.K. (1987). *Farewell to Reason*, Verso.

These are scrutinized by:

Stove, D.C. (1982). *Popper and After: Four Modern Irrationalists*, Pergamon Press.

5 Randomness

O! Many a shaft at random sent,
Finds mark the archer little meant!

Sir Walter Scott, in *Lord of the Isles*

Random Processes

I have used the word 'random' quite freely so far without really giving a definition of what it means. The reason for that is that I really do not know. Turning the vague ideas we have about what it means into rigorous mathematical statements is surprisingly difficult. What I want to do in this chapter is look a little bit deeper into the concept, and look at how it applies (or does not) in both abstract mathematics and in physical systems.

There are many different ways in which a sequence of events could be said to be 'random'. For example, when we claim that the throw of simple six-sided dice is random we probably mean one or the other of two things: one is that each face of the dice has the same probability of appearing, and the other is that each throw of the dice is independent of any subsequent throw. The second of these is the more useful definition. In more general terms, and lapsing into frequentist language for the time being, we can think of a sequence of random variables $X_1, X_2, \ldots$ as being random if each one has the same probability distribution $P(X)$ assigned to it, and all the probabilities in the sequence are independent. This means, for instance, that the joint probability of X_j and X_{j+1} (or any other two members of the sequence) is just given by the product of the individual probabilities. Independence then is the key concept. A dice could be biased, such as if say a 6 were more probable than any other face, but a sequence of throws would still be in some sense random if one throw had no memory of the preceding throw. It does not really matter what the probability distribution is for a sequence to be random in this sense; it just matters that each event is statistically independent of the others in the chain. Each random event could have a discrete set of possible outcomes (such as the dice) or a continuous set (like the reading on a seismometer). One could thus think a random sequence in which the numbers were drawn from a Gaussian distribution (if the range of possible outcomes is continuous) rather than a uniform one, for example.

On the other hand, suppose we imagine a sequence whose starting point was a random number with some probability, but which had perfect memory. Here X_1 would be a randomly-generated number, but each subsequent number would be a copy of this. You probably would not call this random because it would be a string of identical digits. Nevertheless such a sequence still requires a probabilistic description. This is because a random process is a sequence of random variables, and random variables, as I tried to explain in Chapter 4, are curious things. They are 'measure-valued' objects, smeared out over the space of possibilities. If you insist on a frequentist description, a random process is really an ensemble of potential realizations, each one generated from the underlying probability distributions.

Consider a simple example in which we roll a dice three times in sequence. In the case where there is perfect memory the relevant ensemble only contains six possible realizations: 111, 222, 333, 444, 555, and 666. Once you have thrown the first dice, the remaining two throws are fixed. If, on the other hand, the throws are independent, then the space of realizations is much larger: 111, 112, 113, . . . , 211, 212, . . . , etc. Now imagine that instead of three, we are presented with a sequence of hundreds of events. The space of potential realizations is then enormously large. Probabilities have to be defined on this vast range of possibilities.

The mathematical theory of random processes, sometimes called stochastic processes, thus depends on being able to construct joint probabilities of large sequences of random variables, which can be very tricky to say the least. There are, however, some kinds of random processes where the theory is relatively straightforward. One class is when the sequence has no memory at all, which is the case I discussed above. This type of sequence is sometimes called 'white noise', because it has no discernible structure on any scale. Instrumental noise or 'static' in electronic amplifiers tends to have this property; it also tends to have a Gaussian distribution, courtesy of the central limit theorem. Gaussian white noise is a pretty good paradigm for a truly random process, and it is a useful mathematical model for many physical situations.

In what follows let us imagine that we have a time series (sampled at discrete intervals of time) that is modelled by Gaussian white noise: call this G_t. This kind of process can be used to construct more complicated processes with some degree of memory. For example, suppose we have a sequence X_t defined by the iteration scheme:

$$X_t = aX_{t-1} + G_t.$$

The parameter a is a constant that controls the amount of memory in the system: if $a = 0$ then the process X is just the same as G, which has no memory. If $a > 0$ then the larger its value, the more each step depends on the previous one. There is always some degree of randomness, however, as there is always a bit of Gaussian white noise added in at every iteration. The parameter a can also be negative in which case each step reacts against the previous step: the resulting series will be oscillatory rather than smooth. One can extend this general model to include dependence on other steps too, but I will not go into the details here. The type of sequence represented by this simple model is called an autoregressive process, and it is a useful modelling tool for relatively simply time series phenomena. Each step only remembers the previous one in this specific case, which makes it an example of a *Markov process*. These make up one of the few classes of stochastic model for which a complete theory is known.

To characterize the 'memory' in these processes it is useful to generalize the concept of variance introduced in Chapter 2. Suppose we have two random variables X and Y. The variance of X is defined by

$$\mathrm{Var}(X) = E[X^2] - E[X]^2.$$

A similar expression defines the variance of Y. These two quantities give us an idea of the spread of the distributions of X and Y. But they do not tell us if large values of X tend to be accompanied by large values of Y, or vice-versa, or if there is no correlation between the values of X and Y. To quantify this we define the covariance

$$\mathrm{Cov}(X, Y) = E[XY] - E[X]E[Y],$$

where $E[XY]$ is defined over the joint probability of X and Y:

$$E[XY] = \iint P(x, y)xy\, dx\, dy$$

Clearly if the two variables X and Y are independent then $P(X,Y) = P(X)P(Y)$ so $E[XY] = E[X]E[Y]$ and the covariance is consequently zero. On the other hand if they are not independent the covariance tells us something about the strength of the dependence. For example, in Figure 10 we see the masses and heights plotted for a selection of human individuals. Clearly there is some tendency for those of above-average mass also to have above-average height. This can be quantified as a positive covariance between X (mass) and Y (height).

These considerations also apply to random processes if we take the two variables to be elements of the sequence. Suppose $X = X_t$ and $Y = X_{t+s}$, then the covariance becomes the autocovariance of the process, and it measures how strongly the sequence remembers what it was s steps previously. Note that there is no explicit memory of more than one step backwards in the example I gave above, but each step itself remembers each previous one so there is a memory of a memory going back even further. In fact the covariance of this type of process depends on a^s, which decays very quickly as s increases as long as a is less than one. This rapid decay is typical of Markov processes,

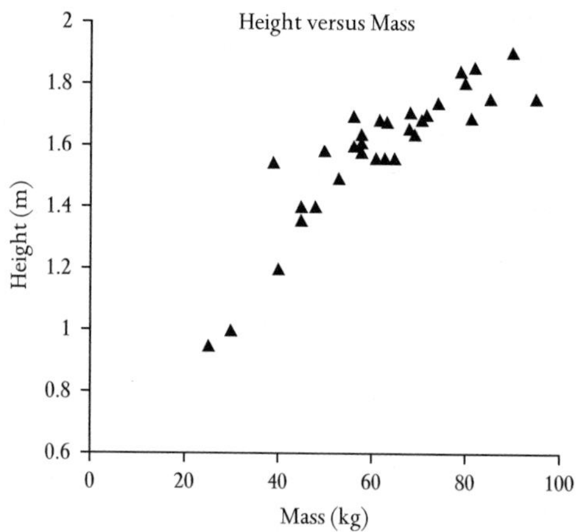

Figure 10 Correlation between height and mass for humans. Tall people tend to be heavier than short ones.

which possess only short-range correlations. If $a = 1$, I get a different kind of process. I will come back to this briefly later on.

This brings me to two important pieces of nomenclature about random processes. One is *stationarity*. In a nutshell a stationary process is one that has no net tendency to wander up or down, that is, no long-term trend. More precisely, in the language of random variables, it is one for which the joint probabilities of different components of the sequence depend only on their *relative* position in the chain. In other words, the joint probabilities of X_t and $Y = X_{t+s}$ depend only on s and not on t. This means that any position in the sequence is statistically equivalent.

The other concept is much more subtle, and it strictly applies only to infinitely long sequences. Roughly speaking, a process is called *ergodic* if a given realization of the sequence visits every part of the probability distribution. In other words, somewhere in an infinitely long realization you will find every possible finite sequence that could be generated from the probability distribution. Not all processes are ergodic, and it is often difficult to prove which ones are.

There is an old story that if you leave a set of monkeys hammering on typewriters for a sufficiently long time then they will eventually reproduce the script for Hamlet. This is not necessarily the case, even if one does allow an infinite time: it depends on the ergodic property applying to their typing. If the monkeys were always to hit two adjoining keys at the same time then they would never produce a script for Hamlet, as the combinations QW or ZX do not appear anywhere in that play!

So far I have discussed idealized stochastic processes based on the mathematical idea of random variables. Surprisingly it is quite easy to generate perfectly deterministic mathematical sequences that behave in much the same fashion, that is, in the way we usually take to characterize indeterministic processes. As a very simple example, consider the following 'iteration' scheme:

$$X_{j+1} = 2X_j \bmod(1).$$

If you are not familiar with the notation, the term mod(1) just means 'drop the integer part'. To illustrate how this works, let us start with a (positive) number, say 0.37. To calculate the next value I double it (getting 0.74) and drop the integer part. Well, 0.74 does not have an integer part so that's fine. This becomes my first iterate. The next one is obtained by putting 0.74 in the formula, that is, doubling it (1.48) and dropping the integer part: result 0.48. Next one is 0.96, and so on. You can carry on this process as long as you like, using each output number as the input state for the following iteration.

Now to simplify things a little bit, notice that, because we drop the integer part each time, all iterates must lie in the range between 0 and 1. Suppose I divide this range into two bins, labelled 'heads' for X less than 1/2 and 'tails' for X greater than or equal to 1/2. In my example above the first value of X is 0.37 which is 'heads'. Next is 0.74 (tails); then 0.48 (heads), 0.96 (tails), and so on.

This sequence now mimics quite accurately the tossing of a fair coin. It produces a pattern of heads and tails with roughly 50% frequency in a long run. It is also difficult to predict the next term in the series given only the classification as 'heads' or 'tails'. However, given the seed number which starts off the process, and of course the algorithm, one could reproduce the entire sequence. It is not random, but looks like it is.

One can think of 'heads' or 'tails' in more general terms, as indicating the '0' or '1' states in the binary representation of a number. This method can therefore be used to generate *any* sequence of digits. In fact algorithms like this one are used in computers for generating what are called pseudorandom numbers. They are not precisely random because computers can only do arithmetic to a finite number of decimal places. This means that only a finite number of possible sequences can be computed, so some repetition is inevitable, but these limitations are not always important in practice.

The ability to generate accurately random numbers in a computer has led to an entirely new way of doing science. Instead of doing real experiments with measuring equipment and the inevitable errors, one can now do numerical experiments with pseudorandom numbers in order to investigate how an experiment might work if we could do it. If we think we know what the result would be, and what kind of noise might arise, we can do a random simulation to discover the likelihood of success with a particular measurement strategy. This is called the 'Monte Carlo' approach, and it is extraordinarily powerful. Observational astronomers and particle physicists use it a great deal in order to plan complex observing programmes and convince the powers that be that their proposal is sufficiently feasible to be allocated time on expensive facilities. In the end there is no substitute for real experiments, but in the meantime the Monte Carlo method can help avoid wasting time on flawed projects:

> in real life mistakes are likely to be irrevocable. Computer simulation, however, makes it economically practical to make mistakes on purpose. (John McLeod and John Osborne, in *Natural Automata and Useful Simulations*)

So is there a way to tell whether a set of numbers is really random ? Consider the following sequence:

1415926535897932384626433832795028841971

Is this a random string of numbers? There does not seem to be a discernible pattern, and each possible digit seems to occur with roughly the same frequency. It does not look like anyone's phone number or bank account. Is that enough to make you think it is random?

Actually this is not at all random. If I had started it with a three and a decimal place you might have cottoned on straight away. '3.1415926 . . . ' are the first few digits in the decimal representation of π. The full representation goes on forever without repeating. This is a sequence that satisfies most naïve definitions of randomness. It does, however, provide something of a hint as to how we might construct an operational definition, that is, one that we can apply in practice to a finite set of numbers.

The key idea originates from the Russian mathematician Andrei Kolmogorov, who wrote the first truly rigorous mathematical work on probability theory in 1933 and made major contributions to the theory of Markov processes like those I discussed above. Kolmogorov's approach was considerably ahead of its time, because it used many concepts that belong to the era of computers. In essence, what he did was to provide a definition of the *complexity* of an N-digit sequence in terms of the smallest amount of computer memory it would take to store a program capable of generating the sequence. Obviously one can always store the sequence itself, which means that there is always a program that occupies about as many bytes of memory as the sequence itself, but some numbers can be generated by codes much shorter than the numbers themselves. For example the sequence

11111111111111111111111111111111111

can be generated by the instruction to 'print "1" 35 times', which can be stored in much less memory than the original string of digits. Such a sequence is therefore said to be algorithmically compressible.

The complexity of a sequence is just the length of the shortest program capable of generating it. If no algorithm can be found that compresses the sequence into a program shorter than itself then it is maximally complex and can suitably be defined as random. This is a very elegant description, and it is in good accord with our intuition. However, it is worth saying that it still does not provide us with a way of testing rigorously whether a sequence is random or not. Randomness means disorder, and disorder can come about in many different ways. If an algorithmic compression can be found then that means the sequence is not random, but if one is not found that may just mean we did not look hard enough. Any test done by a finite human brain will never be sufficient to prove randomness.

Predictability in Principle and Practice

The era of modern physics could be said to have begun in 1687 with the publication by Sir Isaac Newton of his great *Philosophiae Naturalis Principia Mathematica*, (the 'Principia' for short). In this magnificent volume, Newton presented a mathematical theory of all known forms of motion and, for the first time, gave clear definitions of the concepts of force and momentum. Within this general framework he derived a new theory of Universal Gravitation and used it to explain the properties of planetary orbits previously discovered but unexplained by Kepler. The classical laws of motion and his famous 'inverse square law' of gravity have been superseded by more complete theories when dealing with very high speeds or very strong gravity, but they nevertheless continue to supply a very accurate description of our everyday physical world.

Newton's laws have a rigidly *deterministic* structure. What I mean by this is that, given precise information about the state of a system at some time then one can use Newtonian mechanics to calculate the precise state of the system at any later time. The orbits of the planets, the positions of stars in the sky, and the occurrence of eclipses can all be predicted to very high accuracy using this theory.

At this point it is useful to mention that most physicists do not use Newton's laws in the form presented in the Principia, but in a more elegant language named after Sir William Rowan Hamilton. The point about Newton's laws of motion is that they are expressed mathematically as differential equations: they are expressed in terms of rates of changes of things. For instance, the force on a body gives the rate of change of the momentum of the body. Generally speaking, differential equations are very nasty things to solve, which is a shame because a great deal of theoretical physics involves them. Hamilton realized that it was possible to express Newton's laws in a way that did not involve clumsy mathematics of this type. His formalism was equivalent, in the sense that one could obtain the basic differential equations from it, but easier to use in general situations. The key concept he introduced—now called the hamiltonian—is a single mathematical function that depends on both the positions q and momenta p of the particles in a system, say $H(q,p)$. This function is constructed from the different forms of energy (kinetic and potential) in the system, and how they depend on the p's and q's, but the details of how this works out do not matter. Suffice to say that knowing the hamiltonian for a system is tantamount to a full classical description of its behaviour.

Hamilton was a very interesting character. He was born in Dublin in 1805 and showed an astonishing early flair for languages, speaking 13 of them by the time he was 13. He graduated from Trinity College aged 22, at which point he was clearly a whiz-kid at mathematics as well as languages. He was immediately made professor of astronomy at Dublin and Astronomer Royal for Ireland. However, he turned out to be hopeless at the practicalities of observational work. Despite employing three of his sisters to help him in the observatory he never produced much of astronomical interest. Mathematics and alcohol were the two loves of his life.

It is a fascinating historical fact that the development of probability theory during the late seventeenth and early eighteenth century coincided almost exactly with the rise of Newtonian Mechanics. It may seem strange in retrospect that there was no great philosophical conflict between these two great intellectual achievements since they have mutually incompatible views of prediction. Probability applies in unpredictable situations; Newtonian Mechanics says that everything is predictable. The resolution of this conundrum may owe a great deal to Laplace, who contributed greatly to both fields. Laplace, more than any other individual, was responsible for elevating the deterministic world-view of Newton to a scientific principle in its own right. To quote:

> We ought then to regard the present state of the Universe as the effect of its preceding state and as the cause of its succeeding state.

According to Laplace's view, knowledge of the initial conditions pertaining at the instant of creation would be sufficient in order to predict everything that subsequently happened. For him, a probabilistic treatment of phenomena did not conflict with classical theory, but was simply a convenient approach to be taken when the equations of motion were too difficult to be solved exactly. The required probabilities could be derived from the underlying theory, perhaps using the kind of symmetry argument I outlined in the previous chapter. The 'randomizing' devices used in all traditional gambling games—roulette wheels, dice, coins, bingo machines, and so on—are well described by Newtonian mechanics. We call them 'random' because the motions involved are just too complicated to make accurate prediction possible. Nevertheless it is clear that they are just straightforward mechanical devices which are essentially deterministic. On the other hand, we like to think the weather is predictable, at least in principle, but with much less evidence that it is so!

Astronomy provides a nice example that illustrates how easy it is to make things too complicated to solve. Suppose we have two massive bodies orbiting in otherwise empty space. They could be the Earth and Moon, for example, or a binary star system. Each of the bodies exerts a gravitational force on the other that causes it to move. Newton himself showed that the orbit followed by each of the bodies is an ellipse, and that both bodies orbit around their common centre of mass. The Earth is much more massive than the Moon, so the centre of mass of the Earth-Moon system is rather close to the centre of the Earth. Although the Moon appears to do all the moving, the Earth orbits too. If the two bodies have equal masses, they each orbit the mid-point of the line connecting them like two dancers doing a waltz.

Now let us add one more body to the dance. It does not seem like too drastic a complication to do this, but the result is a mathematical disaster. In fact there is no known mathematical solution for the gravitational three-body problem, apart from a few special cases. The same applies to the N-body problem for any N bigger than 2. We cannot solve the equations for systems of gravitating particles except by using numerical techniques and very big computers. We can do this very well these days because computer power is cheap. Computational cosmologists can 'solve' the *N*-body problem for billions of particles, by starting with an input list of positions and velocities of all the particles. From this list the forces on each of them due to all the other particles can be calculated. Each particle is then moved a little according to Newton's laws, thus advancing the system by one time-step. Then the forces are all calculated again and the system inches forward in time. At the end of the calculation, the solution obtained is simply a list of the positions and velocities of each of the particles. If you would like to know what would have happened with a slightly different set of initial conditions you need to run the entire calculation again. There is no elegant formula that can be applied for any input: each laborious calculation is specific to its initial conditions.

But it is not only systems with large numbers of particles that pose problems for predictability. Some deceptively simple systems display extremely erratic behaviour. The theory of these systems is less than 50 years or so old, and it goes under the general title of non-linear dynamics. One of the most important landmarks in this field was a study by two astronomers, Michel Hénon and Carl Heiles in 1964. They were interested in what would happens if you take a system with a known analytical solution and modify it. In the language of hamiltonians, let us assume that H_0 describes a system whose evolution we know exactly and H_1 is some perturbation to it. The hamiltonian of the modified system is thus

$$H = H_0(q_i, p_i) + H_1(q_i, p_i)$$

What Hélnon and Heiles did was to study a system whose unmodified form is very familiar to physicists: the simple harmonic oscillator. This is a system which, when displaced from its equilibrium, experiences a restoring force proportional to the displacement. The hamiltonian description for this system involves a function that is quadratic in both p and q. The solution of this system is well known: the general form is a sinusoidal motion and it is used in the description of all kinds of wave phenomena, swinging pendulums and so on. The case Hénon and Heiles looked at had two degrees of freedom, so that the hamiltonian depends on q_1, q_2, p_1, and p_2. However, the two degrees of freedom are independent, meaning that there is uncoupled motion in the two directions. The amplitude of the oscillations is governed by the total energy of the system, which is a constant of the motion of this system. Other than this, the type of behaviour displayed by this system is very rich, as exemplified by the various Lissajous figures shown in the diagram. Note that all these figures are produced by the same type of dynamical system of equations: the different shapes are consequences of different initial conditions and/or different values of the parameters describing the system.

If the oscillations in each direction have the same frequency then one can get an orbit which is a line or an ellipse. If the frequencies differ then the orbits can be much more complicated, but still pretty. Note that in all these cases the orbit is just a line, that is a one-dimensional part of the two-dimensional space drawn on the paper. More generally, one can think of this system as a point moving in a four-dimensional 'phase space' defined by the coordinates q_1, q_2, p_1, and p_2; taking slices through this space reveals qualitatively similar types of structure for, say, p_2 and q_2 as for p_1 and p_2. The motion of the system is *confined* to a lower-dimensional part of the phase space rather than filling up all the available phase space. In this particular case, because each degree of freedom moves in only one of its two available

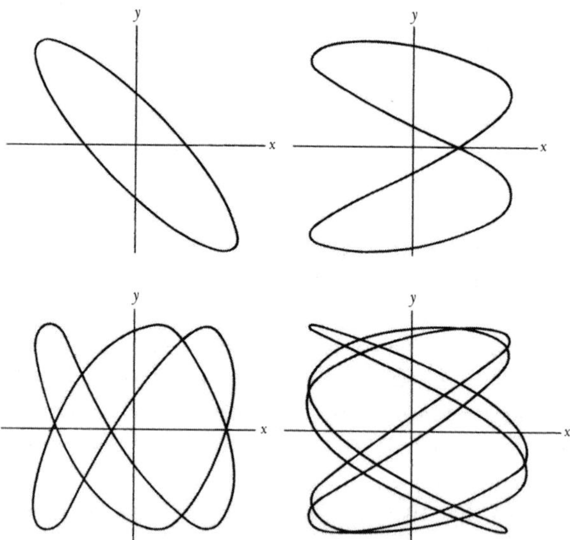

Figure 11 Lissajous figures for a simple harmonic oscillator illustrating the possible forms of 'orbits' for this simple dynamical system.

dimensions, the system as a whole moves in a two-dimensional part of the four-dimensional space.

This all applies to the original, unperturbed system. Hénon and Heiles took this simple model and modified it by adding a term to the hamiltonian that was cubic rather than quadratic and which coupled the two degrees of freedom together. For those of you interested in the details their hamiltonian was of the form

$$H = \frac{1}{2}(p_1^2 + q_1^2 + p_2^2 + q_2^2) + q_1^2 q_2 + \frac{1}{3} q_2^3.$$

The first set of terms in the brackets is the unmodified form, describing a simple harmonic oscillator. The result of this simple alteration is really quite surprising. They found that, for low energies, the system continued to behave like two uncoupled oscillators; the orbits were smooth and well-behaved. This is not surprising because the cubic modifications are smaller than the original quadratic terms if the amplitude is small. For higher energies the motion becomes a bit more complicated, but the phase space behaviour is still characterized by continuous lines, as shown in the top left panel of Figure 12.

However, at higher values of the energy, the cubic terms become more important, and something very striking happens. A two-dimensional slice through the phase space no longer shows the continuous curves that typify the original system, but a seemingly disorganized scattering of dots. It is not possible to discern any pattern in the phase-space structure of this system: to all intents and purposes it is random.

Nowadays we describe the transition from these two types of behaviour as being accompanied by the onset of *chaos*. It is important to note that this system is entirely deterministic, but it generates a phase-space pattern that is quite different from what one would naively expect from the behaviour usually associated with classical hamiltonian systems. To understand how this comes about it is perhaps helpful to think about predictability in classical systems. It is true that precise knowledge of the state of a system allows one to predict its state at some future time. For a single particle this means that precise knowledge of its position and momentum, and knowledge of the relevant H, will allow one to calculate the position and momentum at all future times. But think a moment about what this means. What do we mean by precise knowledge of the particle's position? How precise? How many

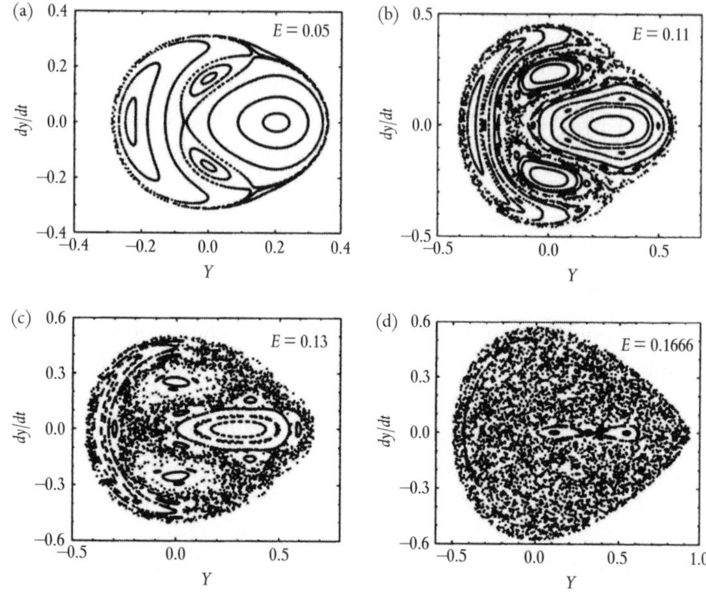

Figure 12 The transition to chaos shown by the Hénon-Heiles system. For small values of the energy E the system behaves smoothly, but this changes drastically as E is increased. Reprinted from *A Modern Approach to Classical Mechanics* by Harald Iro © 2002, with permission from World Scientific Publishing Co Pte Ltd, Singapore.

decimal places? If one has to give the position exactly then that could require an infinite amount of information. Clearly we never have that much information. Everything we know about the physical world has to be coarse-grained to some extent, even if it is only limited by measurement error. Strict determinism in the form advocated by Laplace is clearly a fantasy. Determinism is not the same as predictability.

In 'simple' hamiltonian systems what happens is that two neighbouring phase-space paths separate from each other in a very controlled way as the system evolves. In fact the separation between paths usually grows proportionally to time. The coarse-graining with which the input conditions are specified thus leads to a similar level of coarse-graining in the output state. Effectively the system is predictable, since the uncertainty in the output is not much larger than in the input.

In the chaotic system things are very different. What happens here is that the non-linear interactions represented in the hamiltonian play havoc with the initial coarse-graining. Two phase-space orbits that start out close to each other separate extremely violently as time goes on. In chaotic systems such a divergence is typically an exponential function of time. What happens then is that a tiny change in the initial conditions leads to dramatically different output states; what comes out is practically impossible to predict.

For a real-world illustration of essentially the same phenomenon, look at the picture of cigarette smoke show in Figure 13. Notice that the different tracks of the smoke particles set out very close together on very well defined tracks. This is an example of what is called laminar flow in fluid dynamics. As the particles move further away, much more structure appears and the flow becomes turbulent, in direction analogy with the onset of chaos in the Henon-Heiles system.

Intermezzo: The First Digit Phenomenon

Before going on to talk about some of the more entertaining aspects of random behaviour, I thought it would be fun to introduce a quirky example of how sometimes things that really ought to be random turn out not to be. It is also an excuse to mention yet another astronomer.

Simon Newcomb was born in 1835 in Nova Scotia (Canada). He had no formal education at all, but was a self-taught mathematician and

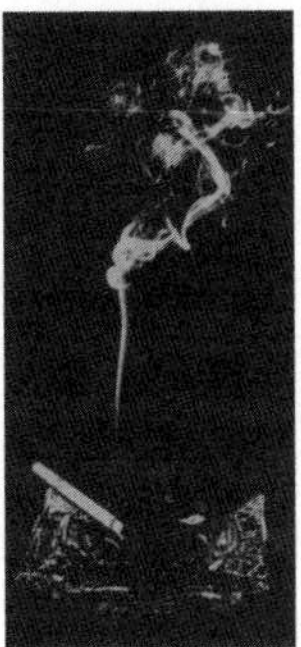

Figure 13 Transition from laminar to turbulent flow in cigarette smoke. Reprinted with permission from *Physics Today*, April 1983, p. 43, Copyright © 1983 American Institute of Physics.

performed astronomical calculations with great diligence. He began work in a lowly position at the Nautical Almanac Office in 1857, and by 1877 he was director. He was professor of Mathematics and Astronomy at Johns Hopkins University from 1884 until 1893 and was made the first ever president of the American Astronomical Society in 1899; he died in 1909.

Newcomb was performing lengthy calculations in an era long before the invention of the pocket calculator or desktop computer. In those days many calculations, including virtually anything involving multiplication, had to be done using logarithms. The logarithm (to the base ten) of a number x is defined to be the number a such that $x = 10^a$. To multiply two numbers whose logarithms are a and b respectively involves simply adding the logarithms: $10^a \times 10^b = 10^{(a+b)}$; adding is a lot easier than multiplying if you have no calculator. The initial logarithms are simply looked up in a table; to find the answer you use the tables again to find the 'inverse' logarithm.

Newcomb was a heavy user of his book of mathematical tables for this type of calculation, and it became very grubby and worn. But Newcomb also noticed that the first pages of the logarithms seemed to have been used much more than the others. This puzzled him greatly. Logarithm tables are presented in order of the first digit of the number required. The first pages contain logarithms for numbers beginning with the digit 1. Newcomb used the tables for a vast range of different calculations of different things. He expected the first digits of numbers that he had to look up to just as likely to be anything. Should not they be randomly distributed? Should not all the pages be equally used?

Once raised, this puzzle faded away until it was re-discovered in 1938 and acquired the name of Benford's law, or the first digit phenomenon. In virtually any list you can think of—street addresses, city populations, lengths of rivers, and so on—there are more entries beginning with the digit '1' than any other digit. In the American Physical Society's list of fundamental constants, no less than 40% begin with the digit 1.

So why is this the case? There is a (relatively) simple answer, and a more complicated one. I will take the simple one first.

Consider street numbers in an address book as an example. Any street will be numbered from 1 to N. It does not really matter what N is as long as it is finite: nobody has ever built an infinitely long street. Now think about the first digits of the addresses. There are 9 possibilities, because we never start an address with 0. On the face of it, we might expect a fraction 1/9 (approximately 11%) of the addresses will start with 1. Suppose N is 200. What fraction actually starts with 1? The answer is more than 50%. Everything from 100 upwards, plus 1, and 11 to 19. Very few start with 9: only 9 itself, and 90–99 inclusive. If N is 300 then there are still more beginning with 1 than any other digit, and there are no more that start with 9. One only gets close to an equal fraction of each starting number if the value of N is an exact power of 10, for example, 1000.

Now you can see why pulling numbers out of an address book leads to a distribution of first digits that is not at all uniform. As long as the numbers are being drawn from a collection of streets each of which has a finite upper limit, then the result is bound to be biased towards low starting digits. Only if every street contained an exact multiple of ten addresses would the result be uniform. Every other possibility favours 1 at the start.

The more complicated version of this argument makes contact with the type of scaling argument I discussed in the previous chapter, and it also is a more suitable explanation for the appearance of this phenomenon in measured physical quantities. Lengths, heights and weights of things are usually measured with respect to some reference quantity. In the absence of any other information, one might imagine that the distribution of whatever is being measured possesses some sort of invariance or symmetry with respect to the scale being chosen. In this case the prior distribution can be taken to have the Jeffreys form, which is uniform in the logarithm. However as before, there obviously must be a cut-off at some point. Suppose that there are many powers of ten involved before this upper limit is reached. In this case the probability that the first digit is D is just given by

$$P(D) = \frac{\int_D^{D+1} P(x)dx}{\int_1^{10} P(x)dx} = \log\left(1 + \frac{1}{D}\right)$$

I have assumed that $P(x)$ is proportional to $1/x$, which integrates to give the logarithm. The shape of this distribution is shown in the Figure. Note that about 30% of the first digits are expected to be 1. Of course I have made a number of simplifying assumptions that are unlikely to be exactly true, but I think this captures the essential reason for the curious behaviour of first digits. If nothing else, it

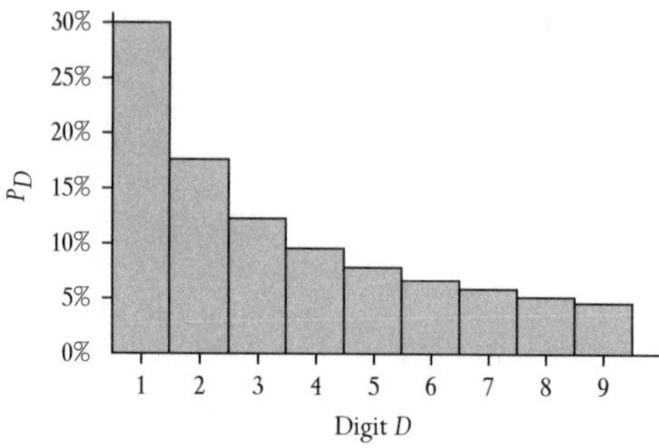

Figure 14 The first-digit phenomenon. The predicted distribution of first-digit frequencies assuming a logarithmic distribution as described in the text.

provides a valuable lesson that you should be careful in what you assume is random!

Points, Patterns, and Poisson

So far in this discussion of randomness I have concentrated on sequences of variables, such as measurements of some variable at successive times. There are, however, many other possible manifestations of randomness. I cannot possibly discuss all of them here, but it is worth introducing a couple of examples to show how randomness can actually appear surprisingly structured.

To start with, we will look at point processes. These are structures in which the random element is a 'dot' that occurs at some location in time or space. Such processes occur in a wide range of contexts: arrival of buses at a bus stop, photons in a detector, darts on a dartboard, and so on.

Let us suppose that we think of such a process happening in time, although what follows can straightforwardly be generalized to things happening over an area (such as a dartboard) or within some higher-dimensional region. The 'most' random way of constructing a point process is to assume that each event happens independently of every other event, and that there is a constant probability per unit time, say λ, of an event happening. This type of process is called a Poisson process, after the French mathematician Siméon–Denis Poisson, who was born in 1781. He was one of the most creative and original physicists of all time: besides fundamental work on electrostatics and the theory of magnetism for which he is famous, he also built greatly upon Laplace's work in probability theory. His principal result was to derive a formula giving the number of random events if the probability of each one is very low. The Poisson distribution, as it is now known and which I will come to shortly, is related to this original calculation; it was subsequently shown that this distribution amounts to a special case of the binomial distribution. Just to add to the connections between probability theory and astronomy, it is worth mentioning that in 1833 Poisson wrote an important paper on the motion of the Moon.

In a finite interval of duration T the mean (or expected) number of events for a Poisson process will obviously just be $\mu = \lambda T$. The full distribution is then

$$P(n) = \frac{\mu^n}{n!}\exp(-\mu).$$

This gives the probability that a finite interval T contains exactly n events. It can be neatly derived from the binomial distribution by dividing the interval into a very large number of very tiny pieces, each one of which becomes a Bernoulli trial. The probability of success (i.e. of an event occurring) in each trial is extremely small, but the number of trials becomes extremely large in such a way that the mean number of successes is μ. In this limit the binomial distribution takes the form of the above expression. The variance of this distribution is interesting: it is also μ. This means that the typical fluctuations within the interval are of order $\mu^{1/2}$ on a mean level of μ. This means the fractional variation is of the famous 'one over root n' form that is a useful estimate of the expected variation in point processes. If football were a Poisson process with a mean number of goals per game of, say 2, then we would expect most games to have 2 plus or minus $\sqrt{2}$ goals, that is, between about 0.6 and 3.4. That is not far from what is observed.

As I mentioned above, this idea can be straightforwardly extended to higher-dimensional processes. If points are scattered over an area with a mean probability per unit area λ then the mean number in a finite area A is just $\mu = \lambda A$ and the same formula applies. As a matter of fact I first learned about the Poisson distribution when I was at school, doing A-level mathematics (which in those days actually included some mathematics). The example used by the teacher to illustrate this particular bit of probability theory was a two-dimensional one from biology. The skin of a fish was divided into little squares of equal area, and the number of parasites found in each square was counted. A histogram of these numbers accurately follows the Poisson form. For years I laboured under the delusion that it was given this name because it was something to do with fish!

This is all very well, but point processes are not always of this Poisson form. Points can be clustered, so that having one point at a given position increases the conditional probability of having others nearby. For example, galaxies are distributed throughout space in a clustered pattern that is far from the Poisson form. The statistical analysis of clustered point patterns is a fascinating subject, because it makes contact with the way in which our eyes and brain perceive pattern. I can only touch on this idea here, but to see what I am talking about look at the two patterns contained in Figure 15.

You will have to take my word for it that one of these is a realization of a two-dimensional Poisson point process and the other contains correlations between the points. I show this example in popular talks and get the audience to vote on which one is random pattern and which one has some pattern. The vast majority usually think that the bottom panel is the one that is random and the top one is the one with structure to it. It is not hard to see why. The bottom pattern has no discernible shape to it, whereas the top one seems to offer a profusion of linear, filamentary features and concentrated clusters.

In fact, the top picture was generated by a Poisson process using the Monte Carlo technique mentioned above. All the structure that is visually apparent is imposed by our own sensory apparatus, which has evolved to be so good at discerning patterns that it finds them when they are not even there! The bottom process is also generated by a Monte Carlo technique, but the algorithm is more complicated. In this case the presence of a point at some location suppresses the probability of having other points in the vicinity. Each event has a zone of avoidance around it; points are anti-correlated. The result of this is that the pattern is much smoother than a truly random process should be.

The tendency to find things that are not there is quite well known to astronomers. The constellations which we all recognize so easily are not physical associations of stars, but are just chance alignments on the sky of things at vastly different distances in space. One could just as easily name features in the random map in Figure 15.

The final type of randomness I would like to mention is one that has a particularly strong resonance in the year 2005. In 1905 Albert Einstein had his 'year of miracles' in which he published three papers that changed the course of physics. One of these is extremely famous: the paper that presented the special theory of relativity. The second was a paper on the photoelectric effect that led to the development of quantum theory, which I discuss in Chapter 7. The third paper is not at all so well known. It was about the theory of Brownian motion. Before describing what this is about it is worth mentioning the poorly

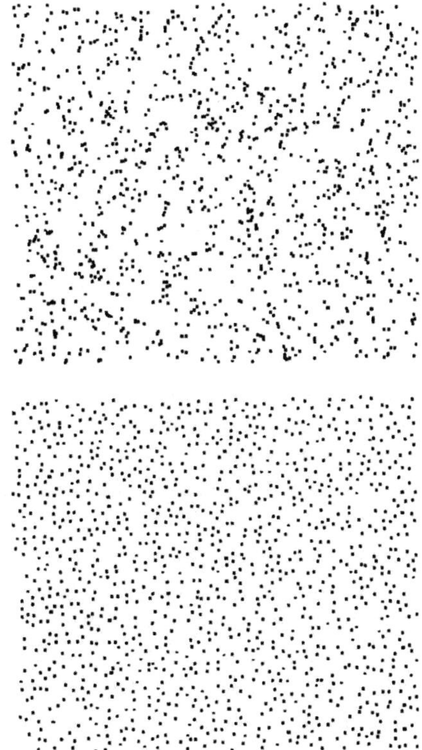

Figure 15 Randomness versus structure in point processes. One of these is random and the other contains structure. Can you tell which is which? From *Bully for Brontosaurus: Reflections in Natural History* by Stephen Jay Gould. Copyright © 1991 by Stephen Jay Gould. Used by Permission of WW Norton & Company, Inc.

recognized fact that Einstein spent an enormous amount of time and energy working on problems in statistical physics.

Brownian motion is the perpetual jittering observed when small particles such as pollen grains are immersed in a fluid. It is now well known that these motions are caused by the constant bombardment of the grain by the fluid molecules. The molecules are too small to be seen directly, but their presence can be inferred from the visible effect on the much larger grain. The mathematical modelling of this process was pioneered by Einstein and Smoluchowski, but has now become a very sophisticated field of mathematics in its own right. I do not want to go into too much detail about the modern approach for fear of getting far too technical, so I will concentrate on the original idea.

Einstein took the view that Brownian motion could be explained in terms of a type of stochastic process called a 'random flight'. I think the first person to formulate this type of phenomenon was the statistician Karl Pearson. The problem he posed concerned the famous drunkard's walk. A man starts from the origin and takes a step of length L in a random direction. After this step he turns through a random angle and takes another step of length L. He repeats this process n times. What is the probability distribution for R, his total distance from the origin after these n steps? Pearson did not actually solve this problem, but posed it in a letter to *Nature* in 1905. A week later, a reply from Lord Rayleigh was published in the same journal. The latter had solved essentially the same problem in a different context way back in 1880.

Pearson's problem is a restricted case of a random walk, with each step having the same length. The more general case allows for a distribution of step lengths as well as random directions. To give a nice example for which virtually everything is known in a statistical sense, consider the case where the components of the step, that is, x and y, are independent Gaussian variables, which have zero mean so that there is no preferred direction:

$$p(x) = \frac{1}{\sigma\sqrt{2\pi}} \exp\left[-\frac{x^2}{2\sigma^2}\right]$$

A similar expression holds for $p(y)$. Now we can think of the entire random walk as being two independent walks in x and y. Incidentally, the autoregressive process I described earlier on is equivalent to such a one-dimensional random walk if $a = 1$. After n steps the total displacement in x, say, x_n is given by

$$p(x_n) = \frac{1}{n\sigma\sqrt{2\pi}} \exp\left[-\frac{x_n^2}{2n\sigma^2}\right]$$

and again there is a similar expression for y_n. Notice that each of these has zero mean. On average, meaning on average over the entire probability distribution of realizations of the walk, the drunkard does not go anywhere. In each individual walk he certainly does go somewhere. The total net displacement from the origin, r_n, is given by Pythagoras' theorem:

$$r_n^2 = x_n^2 + y_n^2$$

from which it is quite easy to find the probability distribution to be

$$p(r_n) = \frac{r_n}{n\sigma^2} \exp\left(-\frac{r_n^2}{2n\sigma^2}\right)$$

This is called the Rayleigh distribution, and this kind of process is called a Rayleigh 'flight'. The mean value of the displacement, $E[r_n]$, is just $\sigma\sqrt{n}$. By virtue of the ubiquitous central limit theorem, this

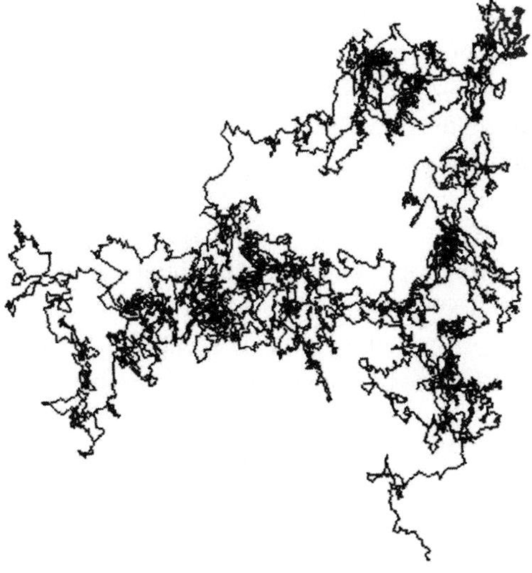

Figure 16 A computer-generated example of a random walk in two dimensions.

result also holds in the original case discussed by Pearson in the limit of very large n.

Figure 16 shows a simulation of a Rayleigh random walk. It is quite a good model for the jiggling motion executed by a Brownian particle. Of course not even the most inebriated boozer will execute a truly random walk. One would expect each step direction to have some memory of the previous one. This gives rise to the idea of a correlated random walk. Such objects can be used to mimic the behaviour of geometric objects that possess some stiffness in their joints, such as proteins or other long molecules.

I mentioned already that the theory of Brownian motion and related stochastic phenomena is now considerably more sophisticated than the simple random flight models I have discussed here. The more general formalism can be used to understand many situations involving diffusion, sedimentation and related phenomena. The ability of these intrinsically random processes to yield surprisingly rich patterns is one of the most fascinating aspects of all physical science.

References and Further Reading

For this chapter, I have borrowed many of the ideas found in:

Ford, John. (1983). How Random is a Coin Toss? *Physics Today*, April.

The classic Henon-Heiles paper can be found in:

Henon, M. and Heiles, C. (1963). The Applicability of the Third Integral of Motion: Some Numerical Examples. *Astronomical Journal*, 69, 73.

A great collection of very important papers on random processes is presented in:

Wax, Nelson. (1954). *Selected Papers on Noise and Stochastic Processes*, Dover.

Einstein's pioneering paper on Brownian Motion can be found in:

Einstein, A. (1905). *Annalen der Physik*, 17, 549.

Random walks are covered in immense detail in:

Hughes, B.D. (1995). *Random Walks and Random Environments*, Oxford University Press.

From Engines to Entropy

Where is the Life we have lost in living?
Where is the wisdom we have lost in knowledge?
Where is the knowledge we have lost in information?

T.S. Eliot, in The Rock

The Laws of Thermodynamics

When I was an undergraduate studying physics my tutor introduced me to thermodynamics by explaining that Ludwig Boltzmann committed suicide in 1906, as did Paul Ehrenfest in 1933. Now it was my turn to study what had driven them both to take their own lives. I did not think this was the kind of introduction likely to inspire a joyful curiosity in the subject, but it probably was not the reason why I found the subject as difficult as I did. It was a hard subject because it seemed to me to possess arbitrary rules that had to be memorized. Lurking somewhere under this set of rules was something statistical, but what it was or how it worked was never made clear. I was frequently told that the best thing to do was just memorize all the different examples given and not try to understand where it all came from. I tried doing this, but partly because I have a very poor memory I did not do very well in the final examination on this topic. I have been prejudiced against it ever since.

It was only after becoming an astrophysicist that I have revisited this subject and come to the conclusion that it is actually a very beautiful one. Unfortunately the features that make it beautiful as well as powerful were never made clear to me as a student. Although I'm by no means an expert, I want in this Chapter to give some idea of why I have changed my mind and how I think thermodynamics should be explained. Consistent with the theme of this book, I will try to argue that it all depends on an understanding of the role of probability.

Classical thermodynamics is essentially a formal system of logic that descends entirely from a set of four axioms, known as the *Laws of Thermodynamics*. What I want to do in this Chapter is to look briefly at the historical development of this subject and show how even the greatest minds of the nineteenth century had enormous problems understanding the relatively simple statistical concepts that underpin it. Sadly, the confused birth of statistical thermodynamics is still reflected in the way in which it is taught in schools and universities. Although I have not got any particularly original insights of my own to offer, I hope that this discussion might at least be an interesting illustration of the labyrinthine processes by which science establishes its conceptual foundations. I do not really want to go into a great deal of formal rigour in discussing the nature of thermodynamic law, so I hope experts will forgive me for merely outlining the essential character of the basic axioms.

Given the emphasis on the topic as a system of logic, it is rather ironic that the first of them is actually called the *Zeroth Law*. This is the simplest to understand because all it does is establish the concept of temperature and its relationship to the idea of thermal equilibrium. Suppose we have a system divided into subsystems. The zeroth law basically says that we can associate a quantity (called the temperature) with each subsystem such that if two subsystems (say A and B) are separately brought into equilibrium with a third subsystem (C), they must then be in equilibrium with each other. If C is a particular type of subsystem called a thermometer then it is clear that this law means that two bodies are in equilibrium with each other if they are at the same temperature. To labour the point even further, it means that if body A is the same temperature as body C and body B is the same temperature as body C then body A is at the same temperature as body B. This sounds awfully obvious, but I did say that it is a *formal* system . . .

The *First Law* is essentially a statement of the law of conservation of energy. To keep things simple I will just assume that there are only two relevant forms of energy: heat and mechanical work. If a system has total energy E and we do something to it that involves either changes in the mechanical work done by the system (W) or its heat content Q then the first law is just that

$$dE = dQ + dW$$

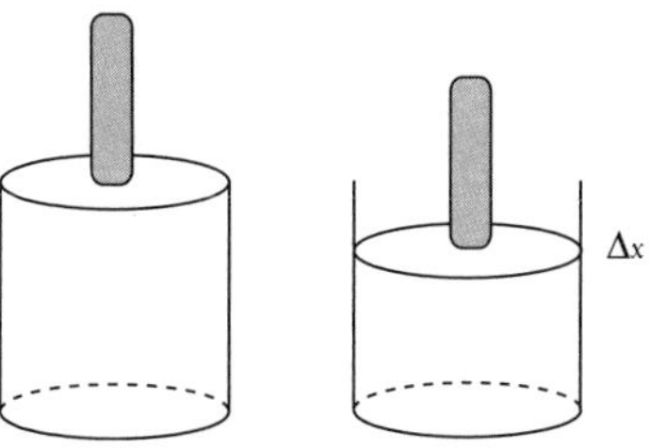

Figure 17 Using a piston to compress gas involves doing work in moving the piston a small distance Δx.

This also seems obvious to modern eyes, but it is worth stressing that it is only relatively recently that it was realized that heat was a form of energy. Until the mid-nineteenth century, heat was described by the caloric theory according to which heat was some kind of 'subtle fluid'. To give an example of how the first law works, imagine a very simple system consisting of a cylinder filled with gas, equipped with a piston at one end. Assume the gas is perfect and the piston slides without friction or anything else that might lead to dissipation of energy. The seal of the piston is also perfect, so that no gas escapes, and the whole thing is isolated, so that no heat flows in or out whatever you do to the piston. Let the piston have cross-sectional area A and the gas inside have pressure P. Now consider what happens if you push the piston very slightly into the cylinder. If it is displaced by an infinitesimally small amount dx then the mechanical work done is just 'force times distance'. The pressure of the gas supplies a force per unit area P so the force opposing the displacement of the piston is just PA. It follows that

$$dE = dW = Fdx = \frac{F}{A}(Adx) = -PdV$$

where dV is the change in volume of the gas contained by the piston. The minus sign appears because the volume of gas decreases when the piston is pushed inwards. This little thought-experiment leads us to an interesting equation for the pressure as the rate of change of the internal energy of the system when the volume is changed but no heat is transferred:

$$\left(\frac{\partial E}{\partial V}\right)_Q = -P$$

Such a change is usually called *adiabatic*.

But what happens if we keep the piston fixed, so that the volume occupied by the gas does not change and the system does no mechanical work, but find some way of injecting energy into it. To understand this situation we have to introduce a new concept, *entropy*. The role and meaning of entropy is the main cause of confusion in the development of thermodynamics, as I will try to explain as we go on. For the time being, however, I will just adopt the following 'macroscopic' definition. If a system is at temperature T when a small amount of heat dQ is added to the system then the change in entropy is

$$dS = \frac{dQ}{T}.$$

You can see that this definition means that when no heat is exchanged with its surroundings ($dQ = 0$), the entropy of the system is constant ($dS = 0$). When heat is exchanged (and everything else is constant) the change in energy is

$$dS = dQ = Tds,$$

which means that

$$\left(\frac{\partial E}{\partial S}\right)_V = T$$

Putting these two things together we get a simple statement of the first law:

$$dE = TdS - PdV$$

This is not the most general form of the first law because there are types of work that might be included other than mechanical (such as the work done moving a charged body in an electrical field). It is also possible to imagine an experiment in which the amount of material in the cylinder (i.e. the number of gas particles) could change. These require additional terms to be added to the preceding equation, but do not alter its meaning.

I have not really described what the thing called entropy actually represents physically because that is not really necessary in this formal exposition. However, it is worth saying that entropy measures something about the state of order in a physical system: it tells us the extent to which energy can be mobilized to perform useful work. If we create a lot of entropy in a process then there is less energy available to use for other things.

Now we are in a position to move onto the famous *Second Law*, which states that the entropy of an isolated system can never decrease. To be a little more precise let us define the finite change in entropy of a system as the result of lots of small changes, that is

$$\Delta S = \int \frac{dQ}{T} \geq 0.$$

This integral must be taken over a series of reversible changes, so that as each tiny dollop of heat is added or subtracted the system is allowed to adjust to a new configuration. In other words the path over which the entropy change is measured must represent a series of equilibrium states of the system. This may be difficult to realize in practice, but remember again that this is a formalized conceptual system. The second law has such a wide range of applicability and has such far-reaching consequences that it occupies a unique place in natural philosophy. For example, in his popular and highly influential book, *The Nature of the Physical World*, the British astrophysicist Arthur Stanley Eddington wrote that:

> The law that entropy always increases—the second law of thermodynamics—holds, I think, the supreme position among the laws of Nature. If someone points out to you that your pet theory of the Universe is in disagreement with Maxwell's equations, then so much the worse for Maxwell's equations. If it is found to be contradicted by observation, well, these experimentalists do bungle things sometimes. But if your theory is found to be against the second law of thermodynamics I can give you no hope; there is nothing for it but to collapse in deepest humiliation.

I will come back to the Second Law, how it emerges from more fundamental considerations and why it is held in such lofty esteem shortly, but first I need to complete the list with the *Third Law*, which is also sometimes called Nernst's theorem. This states that at the absolute zero of temperature the entropy of any system is zero. To put it another way, this law states that no finite series of operations can ever cool a body down from a finite temperature to absolute zero.

Historical Interlude

So far I have presented the field of thermodynamics as a neat and tidy system of axioms and definitions. The resulting laws are written in the language of idealized gases, perfect mechanical devices and reversible equilibrium paths, but despite this have many applications in realistic practical situations. What is interesting about these laws is that it took a very long time to establish them even at this macroscopic level. The deeper understanding of their origin in the microphysics of atoms and molecules took even longer and was an even more difficult journey. I will come to a discussion of the statistical treatment of thermodynamics shortly, but first it is appropriate to celebrate the tangled history of this fascinating subject. Unlike quantum physics and relativity, thermodynamics is not regarded as a very 'glamourous' part of science by the general public, but it did occupy the minds of the greatest physicists of the nineteenth century, and I think the story deserves to be better appreciated. I do not have space to give a complete account, so I apologize in advance for those I have omitted

The story begins with Sadi Carnot, who was born in 1796. His family background was, to say the least, unusual. His father Lasare was known as the 'Organizer of Victory' for the Revolutionary Army in 1794 and subsequently became Napoleon's minister of war. Against all expectations he quit politics in 1807 and became a mathematician. Sadi had a brother, by the splendid name of Hippolyte, who was also a politician and whose son became president of France. Sadi himself was educated partly by his father and partly at the Ecole Polytechnique. He served in the army as an engineer and was eventually promoted to Captain. He left the army in 1828, only to die of cholera in 1832 during an epidemic in Paris.

Carnot's work on the theory of heat engines was astonishingly original and eventually had enormous impact, essentially creating the new science of thermodynamics, but he only published one paper before his untimely death and it attracted little attention during his lifetime. *Reflections on the Motive Power of Fire* appeared in 1824, but its importance was not really recognized until 1849, when it was read by William Thomson (later Lord Kelvin) who, together with Rudolf Clausius, made it more widely known.

In the late eighteenth century, Britain was in the grip of an industrial revolution largely generated by the use of steam power. These engines had been invented by the pragmatic British, but the theory by which they worked was non-existent. Carnot realized that steam-driven devices in use at the time were horrendously inefficient.

As a nationalist, he hoped that by thinking about the underlying principles of heat and energy he might be able to give his native France a competitive edge. He thought about the problem of heat engines in the most general terms possible, even questioning whether there might be an alternative to steam as the best possible 'working substance'. Despite the fact that he employed many outdated concepts, including the so-called caloric theory of heat, Carnot's paper was full of brilliant insights. In particular he considered the behaviour of an idealized friction-free engine in which the working substance moves from a heat source to a heat sink in a series of small equilibrium steps so that the entire process is reversible. The changes of pressure and volume involved in such a process are now known as a Carnot cycle.

By remarkably clear reasoning, Carnot was able to prove a famous theorem that the efficiency of such a cycle depends only on the temperature T_{in} of the heat source and the temperature T_{out}. He showed that the maximum fraction of the heat available to be used to do mechanical work is independent of the working substance and is equal to $(T_{in} - T_{out})/T_{out}$; this is called Carnot's theorem. Carnot's results were probably considered too abstract to be of any use to engineers, but they contain ideas that are linked with the First Law of Thermodynamics, and they eventually led Clausius and Thomson independently to the statement of the Second Law.

Meanwhile, in Manchester, a young man by the name of James Prescott Joule was growing up in a wealthy brewing family. He was born in 1818 and was educated at home by Dalton. He became interested in science and soon started doing experiments in a laboratory near the family brewery. He was a skilful practical physicist and was able to measure the heat and temperature changes involved in various situations. Between 1837 and 1847 he established the basic principle that heat and other forms of energy (such as mechanical work) were equivalent and that, when all forms are included, energy is conserved. Joule measured the amount of mechanical work required to produce a given amount of heat in 1843, by studying the heat released in water by the rotation of paddles powered by falling weights. The SI unit of energy is named in his honour.

William Thomson, First Baron Kelvin of Largs, was born in 1824 and came to dominate British physics throughout the second half of the nineteenth century. He was extremely prolific, writing over 600 research papers and several books. No one since has managed to range so widely and so successfully across the realm of natural sciences. He was also unusually generous with his ideas (perhaps because he had so many), and in giving credit to other scientists, such as Carnot. He was not entirely enlightened, however: he was a vigourous opponent of the admission of women to Cambridge University. Kelvin worked on many theoretical aspects of physics, but was also extremely practical. He directed the first successful transatlantic cable telegraph project, and his house in Glasgow was one of the first to be lit by electricity. Unusually among physicists he became wealthy through his scientific work.

One of the keys to Kelvin's impact on science in Britain was that immediately after graduating from Cambridge in 1845 he went to work in Paris for a year. This opened his eyes to the much more sophisticated mathematical approaches being used by physicists on the continent. British physics, especially at Cambridge, had been held back by an excessive reverence for the work of Newton and the rather cumbersome form of calculus (called 'fluxions') it had inherited from him. Much of Kelvin's work on theoretical topics used the modern calculus which had been developed in mainland Europe. More specifically, it was during this trip to Paris that he heard of the paper by Carnot, although it took him another three years to get his hands on a copy. When he returned from Paris in 1846, the young William Thomson became Professor of Natural Philosophy at Glasgow University, a post he held for an astonishing 53 years.

Initially inspired by Carnot's work, Kelvin became one of the most important figures in the development of the theory of heat. In 1848 he proposed an absolute scale of temperature now known as the Kelvin or thermodynamic scale, which practically corresponds with the Celsius scale except with an offset such that the triple point of water, at 0 °C, is at 273.16 K. He also worked with Joule on experiments concerning heat flow.

At around the same time as Kelvin, another prominent character in the story of thermodynamics was playing his part. Rudolf Clausius was born in 1822. His father was a Prussian pastor and owner of a small school that the young Rudolf attended. He later went to university in Berlin to study history, but switched to science. He was constantly short of money, which meant that it took him quite a long time to graduate but he eventually ended up as a professor of physics, first in Zurich and then later in Wurzburg and Bonn. During the Franco-Prussian war, he and his students set up a volunteer ambulance service and during the course of its operations, Rudolf Clausius was badly wounded.

By the 1850s, thanks largely to the efforts of Kelvin, Carnot's work was widely recognized throughout Europe. Carnot had correctly realized that in a steam engine, heat 'moves' as the steam descends from a higher temperature to a lower one. He, however, envisaged that this heat moved through the engine intact. On the other hand, the work of Joule had established The First Law of Thermodynamics, which states that heat is actually lost in this process, or more precisely heat is converted into mechanical work. Clausius was troubled by the apparent conflict between the views of Carnot and Joule, but eventually realized that they could be reconciled if one could assume that heat does not pass spontaneously from a colder to a hotter body. This was the original statement of what has become known as the Second Law of Thermodynamics. The following year, Kelvin came up with a different expression of essentially the same law. Clausius further developed the idea that heat must tend to dissipate and in 1865 he introduced the term "entropy" in the way I adopted it above, as a measure of the amount of heat gained or lost by a body divided by its absolute temperature. An equivalent statement of the Second Law is that the entropy of an isolated system can never decrease: it can only either increase or remain constant. This principle was intensely controversial at the time, but Kelvin and Maxwell fought vigourously in its defence, and it was eventually accepted into the canon of Natural Law.

So far in this brief historical detour, I have focussed on thermodynamics at a macroscopic level, in the form that eventually emerged as the laws of thermodynamics presented in the previous section. During roughly the same period, however, a parallel story was unfolding that revolved around explaining the macroscopic behaviour of matter in terms of the behaviour of its microscopic components. The goal of this programme was to understand quantitative measures such as temperature and pressure in terms of related quantities describing individual atoms or molecules. I will end this bit of history with a brief description of three of the most important contributors to this strand.

James Clerk Maxwell was probably the greatest physicist of the nineteenth century, and although he is most celebrated for his phenomenal work on the unified theory of electricity and magnetism, he was also a great pioneer in the kinetic theory of gases. He was born in 1831 and went to school at the Edinburgh Academy, which was a difficult experience for him because he had a country accent and invariably wore home-made clothes that made him stand out among the privileged town-dwellers who formed the bulk of the school population. Aged 15, he invented a method of drawing curves using string and drawing pins as a kind of generalization of the well-known technique of drawing an ellipse. This work was published in the Proceedings of the Royal Society of Edinburgh in 1846, a year before Maxwell went to University. After a spell at Edinburgh he went to Cambridge in 1850; while there he won the prestigious Smith's prize in 1854. He subsequently obtained a post in Aberdeen at Marischal College where he married the principal's daughter, but was then made redundant. In 1860 he moved to London but when his father died in 1865 he resigned his post at King's College and became a gentleman farmer doing scientific research in his spare time. In 1874 he was persuaded to move to Cambridge as the first Cavendish Professor of Experimental Physics charged with the responsibility of setting up the now-famous Cavendish laboratory. He contracted cancer five years later and died, aged 48, in 1879.

Maxwell's contributions to the kinetic theory of gases began by building on the idea, originally due to Daniel Bernoulli who we met in Chapter 2, that a gas consists of molecules in constant motion colliding with each other and with the walls of whatever container is holding it. Clausius had already realized that although the gas molecules travel very fast, gases diffuse into each other only very slowly. He deduced, correctly, that molecules must only travel a very short distance between collisions. From about 1860, Maxwell started to work on the application of statistical methods to this general picture. He deduced the probability distribution of molecular velocities in a gas in equilibrium at a given temperature; Boltzmann (see below) independently derived the same result. Maxwell showed how the distribution depends on temperature and also proved that heat must be stored in a gas in the form of kinetic energy of the molecules, thus establishing a microscopic version of the first law of thermodynamics. He went on to explain a host of experimental properties such as viscosity, diffusion, and thermal conductivity using this theory.

Maxwell was fortunate that he was able to make profound intellectual discoveries without apparently suffering from significant mental strain. Unfortunately, the same could not be said of Ludwig Eduard Boltzmann, who was born in 1844 and grew up in the Austrian towns of Linz and Wels, where his father was employed as a tax officer. He received his doctorate from the University of Vienna in 1866 and subsequently held a series of professorial appointments at Graz, Vienna, Munich and Leipzig. Throughout his life he suffered from bouts of depression which worsened when he was subjected to sustained attack from the Vienna school of positivist philosophers, who derided the idea that physical phenomena could be explained in terms of atoms. Despite this antagonism, he taught many students who went on to become very distinguished and he also had a very wide circle of friends. In the end, though, the lack of acceptance of his work got him so depressed that he committed suicide in 1906. Max Planck arranged for his gravestone to be marked with '$S = k \log W$', which is now known as Boltzmann's law and which I will discuss shortly; the constant k is called Boltzmann's constant.

The final member of the cast of characters in this story is Josiah Willard Gibbs. He was born in 1839 and received his doctorate from Yale University in 1863, gaining only the second PhD ever to be awarded in the United States. After touring Europe for a while he returned to Yale in 1871 to become a professor, but he received no salary for the first nine years of this appointment. The university rules at that time only allowed salaries to be paid to staff in need of money; having independent means, Gibbs was apparently not entitled to a salary. Gibbs was a famously terrible teacher and few students could make any sense of his lectures. His research papers are written in a very obscure style which makes it easy to believe he found it difficult to express himself in the lecture theatre. Gibbs actually founded the field of chemical thermodynamics, but few chemists understood his work while he was still alive. His great contribution to statistical mechanics was likewise poorly understood. It was only in the 1890s when his works were translated into German that his achievements were widely recognized. Both Planck and Einstein held him in very high regard, but even they found his work difficult to understand. He died in 1903.

Demons and Arrows: the Rise of Statistical Mechanics

I now want to resume the discussion of thermodynamic concepts, but from the point of view of a microscopic description of matter. This will lead to the resolution of some apparent paradoxes and, I hope, bring out the central importance of the concept of probability for this whole subject. The reason I want to spell this out in detail is partly revenge for the completely garbled way in which I was taught this subject as an undergraduate at Cambridge. When I failed to make sense of my statistical thermodynamics course as a young physics student, I just assumed that it was my fault for being slow on the uptake. Recently I looked at my undergraduate lecture notes again and decided that it was not really my fault at all. I readily admit to being slow on the uptake, but that can be an advantage when what you are asked to take up is nonsense.

A major source of confusion in the way microscopic thermodynamics tends to be taught is a failure to make it clear what probability actually means in the context of statistical mechanics. What I want to do is to draw upon the ideas I have presented in the previous three or four chapters to show that this subject need not be as complicated as it is often presented to be. To do this I will focus on the role and meaning of entropy, because that is usually the source of most confusion,

The best starting point for this exercise is the work of Boltzmann from 1866 where he attempted to derive an expression for entropy using the kinetic theory of gases. The formalism he used is built around the idea of a phase space, of the type I introduced in the previous section when I was talking about orbits. The example of a simple harmonic oscillator I discussed there can be represented as a point moving in a phase space which is two-dimensional. One dimension represents the position of the oscillating object and the other its momentum. At any time, the object has some specific position and some specific velocity; this is represented as a dot in the two-dimensional phase space. As the system evolves this dot moves about this phase plane in a relatively simple way.

One can straightforwardly generalize this idea to describe a particle moving in three dimensions. Now it has three coordinates expressing its position (say x, y, and z) and its momentum s, now a vector with three components $p = (p_x, p_y, p_z)$ representing the velocities in each of the coordinate directions. The phase space is now six-dimensional. This is obviously much harder to visualize than the simple phase plane for a one-dimensional system, although it is conceptually identical. Boltzmann generalized this generalization to the case of N particles each moving in three dimensions, and therefore described by a six-dimensional phase space. Let us define the distribution function as the probability of finding a particle somewhere in this phase space at some time t, that is the probability that a particle is found in a small volume d^3x at position x and with momentum in a small bit of the momentum part of phase space d^3p momentum p. The distribution function is written $\rho(x, p, t)$ Boltzmann defined entropy according to

$$S = -kH = -k \int d^3x d^3p \rho \log \rho;$$

the integral defining the H-function is taken over the whole six-dimensional phase space. I will refer to this as the Boltzmann entropy from now on. This is the point where I got thoroughly confused about statistical mechanics, because I was presented with a 'proof' that the Boltzmann entropy always increases with time. Actually, it is not at all obvious that it does and it's not at all obvious that it is a useful concept in the first place. Boltzmann's idea of introducing probability confuses the distribution function for N particles with something describing N different systems each with one particle. Clearly, he had an attack of frequentists' disease. Boltzmann deserves enormous credit for the profound questions he asked about how, for example, collisions between atoms transfer energy and establish equilibrium states. But he did not quite get there; it was left to Gibbs to produce a more satisfactory theory.

Incidentally, Boltzmann's tombstone is engraved with a different form of the entropy

$$S = k \log W$$

in which W represents the possible number of states of a system with a given total energy. One can see that there is some kind of relationship between these two definitions because the more equivalent states there could be the smaller the probability of finding the system in any particular one. Strangely, however, this definition is not the same as Boltzmann's and it is a bit of a mystery why Planck arranged for it to be put as his epitaph. In any case the more elegant prescription is down to Gibbs.

Gibbs developed his version of statistical mechanics from a different starting point and also introduced an algorithm for setting up descriptions of states in thermodynamic equilibrium. Instead of thinking of a system as N particles each moving in a six-dimensional phase space, Gibbs exploited the device of describing them as a single point moving in a much larger domain. Since each particle requires six dimensions the full phase space for the Gibbs description must possess $6N$ dimensions altogether. Bearing in mind that each particle is an atom and the typical number of atoms in a reasonable everyday amount of gas is Avogadro's Number ($\sim 10^{23}$) then the phase space here has enormous dimensionality. It is nevertheless a beautifully simplifying structure. Gibbs introduced the N-particle distribution ρ_N as the probability of finding the state of the system in a particular part of this super-phase space. This means that ρ_N encodes the probability of finding particle 1 at position x_1I, with momentum p_1, particle 2 at position x_2 with momentum p_2, and so on for all N particles. Each possible position of the system in this phase space is dubbed a *microstate*. Gibbs' definition of entropy is

$$S = -k \int d\tau \rho_N \log \rho_N,$$

where the integral is taken over the whole $6N$-dimensional phase space and $d\tau$ a small element of this space. The Gibbs algorithm for setting up equilibrium states is to maximize the entropy subject to any constraints that might apply. In more common thermodynamical language, this is termed the canonical ensemble. For example, applying this algorithm when the mean energy per particle is fixed and the system is uniform in density leads immediately to the Maxwell–Boltzmann distribution of molecular speeds at a given temperature T:

$$p(v) = 4\left(\frac{m}{2\pi kT}\right)^{\frac{3}{2}} v^2 \exp\left(-\frac{mv^2}{2kT}\right),$$

where m is the molecular mass; the mean kinetic energy is just $3kT$.

This may be ringing vague bells about the way I sneaked in a different definition of entropy in Chapter 4. There I introduced the quantity

$$S = -\int p(x) \log \frac{p(x)}{m(x)} dx$$

as a form of entropy without any reference to thermodynamics at all. One can actually derive the Maxwell–Boltzmann distribution using this form too: the distribution of velocities in each direction is found by maximising the entropy subject to the constraint that the variance is constant (as the variance determines the mean square velocity and hence the mean kinetic energy). This means that the distribution of each component of the velocity must be Gaussian and if the system is statistically isotropic each component must be independent of the others. The speed is thus given by

$$v = \sqrt{v_x^2 + v_y^2 + v_z^2}$$

and each of the components has a Gaussian distribution with variance kT. A straightforward simplification leads to the Maxwell–Boltzmann form.

But what was behind this earlier definition of entropy? In fact the discrete form (with uniform measure)

$$S = -I = -\sum_i p_i \log p_i$$

derives not from physics but from information theory. Claude Shannon derived the expression for the information content I of a probability distribution defined for a discrete distribution in which i runs from 1 to n. Information is sometimes called *negentropy* because in Shannon's definition entropy is simply negative information: the state of maximum entropy is the state of least information. If one uses logarithms to the base 2, the information entropy is equal to the number of yes or no questions required to take our state of knowledge from wherever it is now to one of certainty. If we are certain already we do not need to ask any questions so the entropy is zero. If we are ignorant then we have to ask a lot; our entropy is maximized.

The similarity of this statement of entropy to that involved in the Gibbs algorithm is not a coincidence. It hints at something of great significance, namely that probability enters into the field of statistical mechanics not as a property of a physical system but as a way of encoding the uncertainty in our knowledge of the system. The missing link in this chain of reasoning was supplied in 1965 by the remarkable and much undervalued physicist Ed Jaynes. He showed that if we set up a system according to the Gibbs algorithm, i.e. so that the starting configuration corresponds to the maximum Gibbs entropy, the subsequent evolution of Gibbs entropy is numerically identical to the macroscopic definition given by Clausius I introduced right at the beginning of this Chapter. This is an amazingly beautiful result that is amazingly poorly known.

This interpretation often causes hostility among physicists who use the word 'subjective' to describe its perceived shortcomings. I do not think subjective is really the correct word to use, but there is some sense in which it does apply to thermodynamics. Far from being a shortcoming, I think it is a great strength and I will illustrate a couple of the benefits it brings in the next section.

Arrows and Demons

It is interesting to note that although Maxwell did much to establish the microscopic meaning of the first law of thermodynamics he never really worked on the second law from the same standpoint. Those that did were faced with a conundrum, The behaviour of a system of interacting particles such as the particles of a gas can be expressed in terms of a hamiltonian as I described for simpler examples in the previous chapter. If we have N particles then the appropriate form of the hamiltonian is

$$H(p, q) = \sum_{i=1}^{N} \frac{p_i^2}{2m} + V(\underline{q}_1 \cdots \underline{q}_N);$$

remember that p represents momentum and q position for each particle. The first term represents the kinetic energy and the second is the potential energy involved in the particle–particle interactions. The resulting equations of motion are of the form

$$\dot{q}_i = \frac{\partial H}{\partial p_i} \qquad \dot{p}_i = \frac{\partial H}{\partial q_i}.$$

The dots represent derivatives with respect to time. These equations will be quite complicated because every particle in principle interacts with all the others. However they do possess an important property: everything is reversible. The equations of motion remain the same if one changes the direction of time and changes the direction of motion for all the particles. Consequently, one cannot tell whether a movie of atomic motions is being played forwards or backwards. This means that the Gibbs entropy is actually a constant of the motion: it neither increases nor decreases during Hamiltonian evolution.

But what about the second law of thermodynamics? This tells us that the entropy of a system tends to increase. Our everyday experience tells us this too: we know that physical systems tend to evolve towards states of increased disorder. Heat never passes from a hot body to a cold one. Pour milk into coffee and everything rapidly mixes. How can this directionality in thermodynamics be reconciled with the completely reversible character of microscopic physics?

The answer to this puzzle is surprisingly simple, at least in the framework derived from the Gibbs–Shannon–Jaynes interpretation of entropy. Notice that experimental measurements do not involve atomic properties of matter ('microstates'), but large-scale average things like pressure and temperature ('macrostates'). Appropriate macroscopic quantities are chosen by us as useful things to use because they allow robust repeatable experiments to be performed. By definition, however, they involve a substantial coarse-graining of our description of the system.

Suppose we perform an idealized experiment that starts from some initial macrostate. In general this will generally be consistent with a number—probably a *very* large number—of initial microstates. As the experiment continues the system evolves along a Hamiltonian path so that the initial microstate will evolve into a definite final microstate. This is perfectly symmetrical and reversible. But the point is that we can never have enough information to predict exactly where in the final phase space the system will end up. I touched on this in Chapter 5: determinism does not necessarily allow predictability. If we choose macrovariables so that our experiments are reproducible it is inevitable that the set of microstates consistent with the final macrostate must be larger than the set of microstates consistent with the initial macrostate in any realistic system. Our lack of knowledge means that the probability distribution of the final state is smeared out over a larger phase space volume at the end than at the start. The entropy has increased, not because of anything happening at the microscopic level but because our definition of macrovariables requires it.

This is illustrated in Figure 18. Each individual microstate in the initial collection evolves into one state in the final collection: the narrow

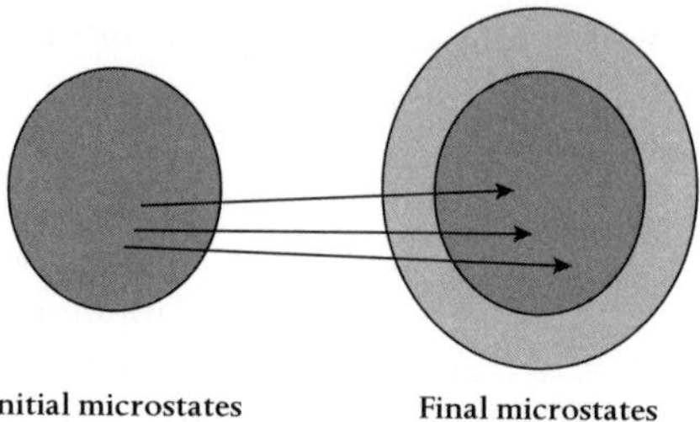

Figure 18 The set of final microstates is never smaller than the initial set because this would make it impossible for an experiment to be repeatable in the macrovariables.

arrows represent hamiltonian evolution. However given only a finite amount of information about the initial state these trajectories must be smeared out. This requires the set of final microstates to acquire a 'buffer zone' around the strictly hamiltonian core. This is the only way to ensure that measurements on such systems will be reproducible.

The 'theoritical' Gibbs entropy remains exactly constant during this kind of evolution, and it is precisely this property that requires the experimental entropy to increase. There is no microscopic explanation of the 2nd law: it arises from our attempt to shoe-horn microscopic behaviour into framework furnished by macroscopic experiments.

Another, perhaps even more compelling demonstration of the so-called subjective nature of probability (and hence entropy) is furnished by Maxwell's demon. This little imp first made its appearance in 1867 or thereabouts and subsequently led a very colourful and influential life. The idea is extremely simple: imagine we have a box divided into two partitions, *A* and *B*. The wall dividing the two sections contains a tiny door which can be opened and closed by a 'demon'—a microscopic being 'whose faculties are so sharpened that he can follow every molecule in its course'. The demon wishes to play havoc with the second law of thermodynamics so he looks out for particularly fast moving molecules in partition *A* and opens the door to allow them (and only them) to pass into partition *B*. He does the opposite thing with partition *B*, looking out for particularly sluggish molecules and opening the door to let them into partition *A* when they approach.

The net result of the demon's work is that the fast-moving particles from *A* are preferentially moved into *B* and the slower particles from *B* are gradually moved into *A*. The net result is that the average kinetic energy of *A* molecules steadily decreases while that of *B* molecules increases. In effect, heat is transferred from a cold body to a hot body, something that is forbidden by the second law.

All this talk of demons probably makes this sound rather frivolous, but it is a serious paradox that puzzled many great minds. Until it was resolved in 1929 by Leo Szilard. He showed that the second law of thermodynamics would not actually be violated if entropy of the entire system (i.e. box + demon) increased by an amount $\Delta S = k\log 2$ every time the demon measured the speed of a molecule so he could decide whether to let it out from one side of the box into the other. This amount of entropy is precisely enough to balance the apparent decrease in entropy caused by the gradual migration of fast molecules from *A* into *B*. This illustrates very clearly that there is a real connection between the demon's state of knowledge and the physical entropy of the system.

By now it should be clear why there is some sense of the word subjective that does apply to entropy. It is not subjective in the sense that anyone can choose entropy to mean whatever he likes, but it is subjective in the sense that it is something to do with the way we manage our knowledge about nature rather than about nature itself. I know from experience that many physicists feel very uncomfortable about the idea that entropy might be subjective even in this sense.

I have to say I feel completely comfortable about the notion: I even think it's obvious.

To see why, consider the example I gave above about pouring milk into coffee. We are all used to the idea that the nice swirly pattern you get when you first pour the milk in is a state of relatively low entropy. The parts of the phase space of the coffee + milk system that contain such nice separations of black and white are few and far between. It's much more likely that the system will end up as a 'mixed' state. But then how well mixed the coffee is depends on your ability to resolve the size of the milk droplets. An observer with good eyesight would see less mixing than one with poor eyesight. In this case entropy, like beauty, is definitely in the eye of the beholder.

References and Further Reading

A much more thorough account of the history of thermodynamics than I have been able to give here is presented in:

Grandy, W.T. (1987). *Foundations of Statistical Mechanics*, Reidel, Dordrecht.

The following are two excellent vigourous polemics about the meaning of entropy. I have drawn substantially from both of them, but they are much more detailed than the brief sketch I have presented here:

Garrett, Anthony J.M. (1991). Macroirreversibility and Microreversibility Reconciled: The Second Law, in *Maximum Entropy in Action*, eds Buck B. and Macaulay V.A., pp. 139–170, Oxford University Press.

Gull, Steven F. (1991). Some Misconceptions about Entropy, in *Maximum Entropy in Action*, eds Buck B. and Macaulay V.A., pp. 171–186, Oxford University Press.

The following two papers are absolute classics of lucidity:

Jaynes, Ed. (1965). Gibbs vs Boltzmann Entropies. *American Journal of Physics*, 33, 391–398.

Shannon, Claude. (1948). A Mathematical Theory of Communication, *Bell Systems Technical Journal*, 27, 379–423 and 623–659.

7

Quantum Roulette

> I think it is safe to say that no one understands quantum mechanics.
>
> Richard Feynman

The Birth of the Quantum

The development of kinetic theory in the latter part of the nineteenth Century represented the culmination of a mechanistic approach to natural philosophy that had begun with Newton two centuries earlier. So successful had this programme been by the turn of the twentieth century that it was a fairly common view among scientists of the time that there was virtually nothing important left to be 'discovered' in the realm of natural philosophy. All that remained were a few bits and pieces to be tidied up, but nothing could possibly shake the foundations of Newtonian mechanics.

But shake they certainly did. In 1905 the young Albert Einstein—surely the greatest physicist of the twentieth century, if not of all time—single-handedly overthrew the underlying basis of Newton's world with the introduction of his special theory of relativity. Although it took some time before this theory was tested experimentally and gained widespread acceptance, it blew an enormous hole in the mechanistic conception of the Universe by drastically changing the conceptual underpinning of Newtonian physics. Out were the 'commonsense' notions of absolute space and absolute time, and in was a more complex 'space-time' whose measurable aspects depended on the frame of reference of the observer.

Relativity, however, was only half the story. Another, perhaps even more radical shake-up was also in train at the same time. Although Einstein played an important role in this advance too, it led to a theory he was never comfortable with: quantum mechanics. A hundred years on, the full implications of this view of nature are still far from understood, so maybe Einstein was correct to be uneasy.

The birth of quantum mechanics partly arose from the developments of kinetic theory and statistical mechanics that I discussed briefly in the previous Chapter. Inspired by such luminaries as Maxwell and Boltzmann, physicists had inexorably increased the range of phenomena that could be brought within the descriptive framework furnished by Newtonian mechanics and the new modes of statistical analysis that they had founded. Maxwell had also been responsible for another major development in theoretical physics: the unification of electricity and magnetism into a single system known as electromagnetism. Out of this mathematical *tour de force* came the realization that light was a form of electromagnetic wave, an oscillation of electric and magnetic fields through apparently empty space. Optical light forms just part of the possible spectrum of electromagnetic radiation, which ranges from very long wavelength radio waves at one end to extremely short wave gamma rays at the other.

With Maxwell's theory in hand, it became possible to think about how atoms and molecules might exchange energy and reach equilibrium states not just with each other, but with light. Everyday experience that hot things tend to give off radiation and a number of experiments—by Wien and others—had shown that there were well-defined rules that determined what type of radiation (i.e. what wavelength) and how much of it were given off by a body held at a certain temperature. In a nutshell, hotter bodies give off more radiation (proportional to the fourth power of their temperature), and the peak wavelength is shorter for hotter bodies. At room temperature bodies give off infra-red radiation, stars have surface temperatures measured in thousands of degrees so they give off predominantly optical and ultraviolet light. In the next Chapter we will see that our Universe is suffused with microwave radiation corresponding to just a few degrees above absolute zero.

The name given to a body in thermal equilibrium with a bath of radiation is a 'black body', not because it is black—the Sun is quite a good example of a black body and it is not black at all—but because it is simultaneously a perfect absorber and perfect emitter of radiation. In other words, it is a body which is in perfect thermal contact with the light it emits. Surely it would be straightforward to apply classical statistical reasoning to a black body at some temperature?

It did indeed turn out to be straightforward, but the result was a catastrophe. One can see the nature of the disaster very straightforwardly by taking a simple idea from classical kinetic theory. In many circumstances there is a 'rule of thumb' that applies to systems in thermal equilibrium. Roughly speaking, the idea is that energy becomes divided equally between every possible 'degree of freedom' the system possesses. For example, if a box of gas consists of particles that can move in three dimensions then, on average, each component of the velocity of a particle will carry the same amount of kinetic energy. Molecules are able to rotate and vibrate as well as move about inside the box, and the equipartition rule can apply to these modes too.

Maxwell had shown that light was essentially a kind of vibration, so it appeared obvious that what one had to do was to assign the same amount of energy to each possible vibrational degree of freedom of the ambient electromagnetic field. Lord Rayleigh and Sir James Jeans did this calculation and found that the amount of energy radiated by a black body as a function of wavelength should vary inversely as the fourth power of the wavelength, as shown in the diagram

Even without doing any detailed experiments it is clear that this is just nonsense. The Rayleigh-Jeans law predicts that even very cold bodies should produce infinite amounts of radiation at infinitely short wavelengths, that is, in the ultraviolet. It also predicts that the total amount of radiation—the area under the curve in the above

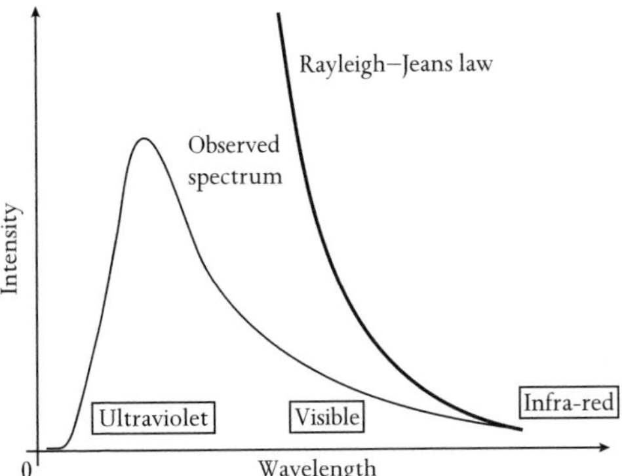

Figure 19 The ultraviolet catastrophe. Attempts to apply statistical theory to radiation resulted in the prediction that bodies should radiate with infinite intensity at very short wavelengths, in contrast to experimental measurements.

figure—is infinite. Even a very cold body should emit infinitely intense electromagnetic radiation. Experiments show that the Rayleigh–Jeans law does work at very long wavelengths but in reality the radiation reaches a maximum (at a wavelength that depends on the temperature) and then declines at short wavelengths. Clearly something is very badly wrong with the reasoning here, although it works so well for atoms and molecules.

It would not be accurate to say that physicists all stopped in their tracks because of this difficulty. It is amazing the extent to which people are able to carry on despite the presence of obvious flaws in their theory. It takes a great mind to realize when everyone else is on the wrong track, and a considerable time for revolutionary changes to become accepted. In the meantime, the run-of-the-mill scientist tends to carry on regardless.

The resolution of this particular fundamental conundrum is accredited to Karl Ernst Ludwig 'Max' Planck, who was born in 1858. He was the son of a law professor, and himself went to university at Berlin and Munich, receiving his doctorate in 1880. He became professor at Kiel in 1885, and moved to Berlin in 1888. In 1930 he became president of the Kaiser Wilhelm Institute, but resigned in 1937 in protest at the behaviour of the Nazis towards Jewish scientists. His life was blighted by family tragedies: his second son died in the First World War; both daughters died in childbirth; and his first son was executed in 1944 for his part in a plot to assassinate Adolf Hitler. After the Second World War the institute was named the Max Planck Institute, and Planck was reappointed director, He died in 1947; by then such a famous scientist that his likeness appeared on the two Deutschmark coin issued in 1958.

Planck had taken some ideas from Boltzmann's work but applied them in a radically new way. The essence of his reasoning was that the ultraviolet catastrophe basically arises because Maxwell's electromagnetic field is a continuous thing and, as such, appears to have an infinite variety of ways in which it can absorb energy. When you are allowed to store energy in whatever way you like in all these modes, and add them all together you get an infinite power output. But what if there was some fundamental limitation in the way that an atom could exchange energy with the radiation field? If such a transfer can only occur in discrete lumps or *quanta*—rather like 'atoms' of radiation—then one could eliminate the ultraviolet catastrophe at a stroke. Planck's genius was to realize this, and the formula he proposed contains a constant that still bears his name. The energy of a light quantum E is related to its frequency ν via $E = h\nu$, where h is Planck's constant: one of the fundamental quantities on which modern physics is based.

Boltzman had shown that if a system possesses two discrete 'states' (say 0 and 1) separated by energy E then at a given temperature the likely relative occupation of the two states is determined by a 'Boltzmann factor' of the form:

$$\frac{n_1}{n_0} = \exp\left(-\frac{E}{kT}\right),$$

so that the higher energy state is exponentially less probable than the lower energy state if the energy difference is much larger than the typical thermal energy kT. On the other hand, if the states are very close in energy compared to the thermal level then they will be roughly equally populated in accordance with the 'equipartition' idea I mentioned above. The trouble with a classical electromagnetic field is that it appears to be able to store infinite energy in short wavelength oscillations by putting a little bit of energy in each of a lot of modes in such a way that the total is divergent. Planck realized that his idea would mean ultra-violet radiation could only be emitted in very energetic quanta, rather than in lots of little bits. Building on Boltzmann's reasoning, he deduced the probability of exciting a quantum with very high energy is exponentially suppressed. This in turn leads to an exponential cut-off in the black-body curve at short wavelengths. Triumphantly, he was able to calculate the exact form of the black-body curve expected in his theory: it matches the Rayleigh–Jeans form at long wavelengths, but turns over and decreases at short wavelengths just as the measurements require. The theoretical Planck curve matches measurements perfectly over the entire range of wavelengths that experiments have been able to probe.

Curiously perhaps, Planck stopped short of the modern interpretation of this: that light (and other electromagnetic radiation) is composed of particles which we now call photons. He was still wedded to Maxwell's description of light as a wave phenomenon, so he preferred to think of the *exchange* of energy as being quantized rather than the radiation itself. Einstein's work on the photoelectric effect in 1905 further vindicated Planck, but also demonstrated that light travelled in packets. After Planck's work, and the development of the quantum theory of the atom pioneered by Niels Bohr, quantum theory really began to take hold of the physics community and eventually it became acceptable to conceive of not just photons but all matter as being part particle and part wave. Photons are examples of a kind of particle known as a boson, and the atomic constituents such as electrons and protons are fermions. (This classification arises from their spin: bosons have spin which is an integer multiple of Planck's constant, whereas fermions have half-integral spin.)

You might have expected that the radical step made by Planck would immediately have led to a drastic overhaul of the system of thermodynamics put in place in the preceding half-a-century, but you would be wrong. In many ways the realization that discrete energy levels were involved in the microscopic description of matter if anything made thermodynamics easier to understand and apply. The point is one that I made in Chapter 2: probabilistic reasoning is usually most difficult when the space of possibilities is complicated. In quantum theory one always deals fundamentally with a discrete space of possible outcomes. Counting discrete things is not always easy, but it is usually easier than counting continuous things.

Much of modern physics research lies in the arena of condensed matter physics, which deals with the properties of solids and gases, often at very low temperatures where quantum effects become important. The statistical thermodynamics of these systems is based on a very slight modification of Boltzmann's result:

$$n_q = \frac{1}{\exp(E_q/kT) \pm 1},$$

which gives the equilibrium occupation of states at an energy level E_q; the difference between bosons and fermions manifests itself as the sign in the denominator. Fermions take the upper 'plus' sign, and the resulting statistical framework is based on the so-called Fermi–Dirac distribution; bosons have the minus sign and obey Bose–Einstein statistics. This modification of the classical theory of Maxwell and Boltzmann is simple, but leads to a range of fascinating phenomena, from neutron stars to superconductivity. Even more amazingly it turns out that the Gibbs entropy discussed in the previous Chapter carries directly over to quantum mechanical systems when expressed in terms of a density matrix.

Waves and Particles

Having argued that, in some ways, quantum theory actually makes physics easier to understand, I now want to explain why you should be very confused by it. When I was an undergraduate I was often told by lecturers that I should find it very difficult, because it is unlike the classical physics I had learned about up to that point. The difference—so I was informed—was that classical systems were predictable, but quantum systems were not. For that reason the microscopic world could only be described in terms of probabilities. I was a bit confused by this, because I already knew that many classical systems were predictable in principle, but not really in practice. I discussed some examples in Chapter 5. It was only when I had studied theory for a long time—almost three years—that I realized what was the correct way to be confused about it. In short quantum probability is a very strange kind of probability that displays many peculiar properties that one doesn't see in the normal 'random' systems like coin-tossing or roulette wheels. Although Einstein was one of the fathers of quantum theory, he detested the idea that there was some fundamental unpredictability in the way nature works. 'God does not play dice with the Universe', he famously remarked. As we shall see it appears that God not only plays dice, but he also cheats.

To see how curious the quantum universe is we have to look at the basic theory. There are different ways of constructing the theory, but I will look at the 'wave' form that is most often taught in introductory books. Incidentally, even the founder of wave mechanics, Erwin Schrödinger, shared Einstein's dislike for a probabilistic interpretation of quantum theory and continually argued against it. Schrödinger was born in 1887 into an Austrian family made rich by a successful oilcloth business run by his father. He was educated at home by a private tutor before going to the University of Vienna where he obtained his doctorate in 1910. During the First World War he served in the artillery, but was posted to an isolated fort where he found lots of time to read about physics. After the end of hostilities he travelled around Europe and started a series of inspired papers on the subject now known as wave mechanics; his first work on this topic appeared in 1926. He succeeded Planck as Professor of Theoretical Physics in Berlin, but left for Oxford when Hitler took control of Germany in

1933. He left Oxford in 1936 to return to Austria but fled when the Nazis seized the country and he ended up in Dublin, at the Institute for Advanced Studies which was created especially for him by the Irish Taoiseach, Eamon de Valera. He remained there happily for 17 years before returning to his native land at the University of Vienna. Sadly, he became ill shortly after arriving there and died in 1961.

Schrödinger was a friendly and informal man who got on extremely well with colleagues and students alike. He was also a bit scruffy even to the extent that he sometimes had trouble getting into major scientific conferences, such as the Solvay conferences which are exclusively arranged for winners of the Nobel Prize. Physicists have never been noted for their sartorial elegance, but Schrödinger must have been an extreme case.

The theory of wave mechanics arose from work published in 1924 by de Broglie who had suggested that every particle has a wave somehow associated with it, and the overall behaviour of a system resulted from some combination of its particle-like and wave-like properties. What Schrödinger did was to write down an equation, involving a hamiltonian describing particle motion of the form I have discussed in previous chapters, but written in such a way as to resemble the equation used to describe wave phenomena throughout physics. The resulting mathematical form for a single particle is

$$i\hbar\frac{\partial\psi}{\partial t} = H\psi = -\frac{\hbar^2}{2m}\nabla^2\psi + V\psi,$$

in which the term ψ is the *wave-function* of the particle. As usual, the hamiltonian H consists of two parts: one describes the kinetic energy (the first term on the right hand side) and the second its potential energy represented by V. This equation—Schrödinger equation—is one of the most important in all physics.

At the time Schrödinger was developing his theory of wave mechanics it had a rival, called matrix mechanics, developed by Heisenberg and others. Paul Dirac later proved that wave mechanics and matrix mechanics were mathematically equivalent; these days physicists generally use whichever of these two approaches is most convenient for particular problems.

Schrödinger's equation is important historically because it brought together lots of bits and pieces of ideas connected with quantum theory into a single coherent descriptive framework. For example, in 1911 Niels

Bohr had begun looking at a simple theory for the hydrogen atom which involved a nucleus consisting of a positively charged proton with a negatively charged electron moving around it in a circular orbit. According to standard electromagnetic theory this picture has a flaw in it: the electron is accelerating and consequently should radiate energy. The orbit of the electron should therefore decay rather quickly. Bohr hypothesized that special states of this system were actually stable; these states were ones in which the orbital angular momentum of the electron was an integer multiple of Planck's constant. This simple idea endows the hydrogen atom with a discrete set of energy levels which, as Bohr showed in 1913, were consistent with the appearance of sharp lines in the spectrum of light emitted by hydrogen gas when it is excited by, for example, an electrical discharge. The calculated positions of these lines were in good agreement with measurements made by Rydberg so the Bohr theory was in good shape. But where did the quantized angular momentum come from?

The Schrödinger equation describes some form of wave; its solutions $\psi(x,t)$ are generally oscillating functions of position and time. If we want it to describe a stable state then we need to have something which does not vary with time, so we proceed by setting the left-hand-side of the equation to zero. The hydrogen atom is a bit like a solar system with only one planet going around a star so we have circular symmetry which simplifies things a lot. The solutions we get are waves, and the mathematical task is to find waves that fit along a circular orbit just like standing waves on a circular string. Immediately we see why the solution must be quantized. To exist on a circle the wave cannot just have any wavelength; it has to fit into the circumference of the circle in such a way that it winds up at the same value after a round trip. In Schrödinger's theory the quantization of orbits is not just an ad hoc assumption, it emerges naturally from the wave-like nature of the solutions to his equation.

The Schrödinger equation can be applied successfully to systems which are much more complicated than the hydrogen atom, such as complex atoms with many electrons orbiting the nucleus and interacting with each other. In this context, this description is the basis of most work in theoretical chemistry. But it also poses very deep conceptual challenges, chiefly about how the notion of a 'particle' relates to the 'wave' that somehow accompanies it. To illustrate the riddle, consider a very simple experiment where particles of some type (say electrons, but it does not

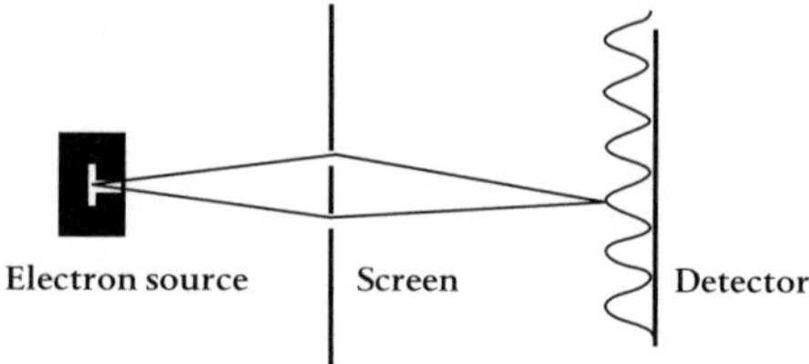

Figure 20 A classic 'two slit' experiment. Somehow electrons managed to travel through both slits at once, interfering with themselves on the way to produce a diffraction pattern at the detector.

really matter; similar experiments can be done with photons or other particles) emerge from the source on the left, pass through the slits in the middle and are detected in the screen at the right.

In a purely 'particle' description we would think of the electrons as little billiard balls being fired from the source. Each one then travels along a well-defined path, somehow interacts with the screen and ends up in some position on the detector. In a 'wave' description we would imagine a wave front emerging from the source, being diffracted by the screen and ending up as some kind of interference pattern at the detector. In quantum theory we have to think of the system as being both. This is forced on us by the fact that we actually observe a pattern of 'fringes' at the detector, indicating wave-like interference, but we also can detect the arrival of individual electrons as little dots. Somehow the propensity of electrons to arrive in positions on the screen is controlled by an element of waviness, but they manage to retain some aspect of their particleness. Moreover, one can turn the source intensity down to a level where there is only every one electron in the experiment at any time. One sees the dots arrive one by one on the detector, but adding them up over a long time still yields a pattern of fringes. Curiouser and curiouser, said Alice.

Eventually the community of physicists settled on a party line that most still stick to: that the wave-function controls the probability of finding an electron at some position when a measurement is made. In fact the mathematical description of wave phenomena favoured by physicists involves complex numbers, so at each point in space at time ψ is a complex number of the form $a + ib$, where i is $\sqrt{(-1)}$; the corresponding probability is given by $|\psi|^2$ which is $a^2 + b^2$. This protocol, however, forbids one to say anything about the state of the particle before it measured. It is delocalized, only possessing a probability to be anywhere in the experiment. One cannot even say which of the two slits it passes through. Somehow, it passes through both or at least some of its wave-function does.

The unspecified nature of the position of the electron extends also to its other properties. For example, being a fermion, an electron possesses spin. One is tempted to think of it as a little cricket ball that can be rotating clockwise or anti-clockwise as it approaches the batsman. But quantum spin is not really like classical spin: batting would be even more difficult if quantum bowlers were allowed! Electron spin is quantized, so that it always has a magnitude which is $\pm 1/2$ (in units of Planck's constant). Until one makes a measurement the state of the system is not specified: it will be either up or down with a 50% probability of each. We could write this as

$$|\psi\rangle \frac{1}{\sqrt{2}}|\uparrow\rangle \pm \frac{1}{\sqrt{2}}|\downarrow\rangle.$$

This gives me an excuse to introduce the rather beautiful 'bra-ket' notation for the state of a quantum system, originally due to Paul Dirac. The two possibilities are 'up' ($\uparrow$) and 'down' ($\downarrow$) and they are contained within a 'ket' which is just a shorthand for a wavefunction describing that particular aspect of the system. A 'bra' would be of the form $\langle\uparrow|$; for the mathematicians this is essentially the complex conjugate of a ket. The two coefficients are there to insure that the total probability of the spin being either up or down is 1, remembering that the probability is the square of the wavefunction. When we make a measurement we will get one of these two outcomes, with a 50% probability of each. At the point of measurement the state changes: if we get 'up' it becomes $|\uparrow\rangle$and if we get 'down' it becomes $|\uparrow\rangle$. Either way, the quantum state of the system has changed from a 'superposition' to a 'pure' state. This means that all subsequent measurements of the spin in this direction will give the same result: the wave-function has 'collapsed' into one particular state.

Incidentally, the general term for a two-state quantum system like this is a qubit, and it is the basis of the tentative steps that have been taken towards the construction of a quantum computer.

Notice that what is essential about this is the role of measurement: the Schrödinger equation itself describes purely Hamiltonian evolution of the wave-function. There is no real difference between this and the classical processes I have described in the preceding Chapters. The collapse of ψ does seem to be an irreversible process, but this is not represented in the equation itself. An extra level of interpretation is needed because we are unable to write down a wave-function that sensibly describes the system plus the measuring apparatus in a single form.

So far this all seems rather similar to the state of a fair coin: it has a 50–50 chance of being heads or tails, but the doubt is resolved when its state is actually observed. Thereafter we know for sure what it is. But this resemblance is only superficial. A coin only has heads or tails, but the spin of an electron does not have to be just up or down. We could rotate our measuring apparatus by 90° and measure the spin to the left (←) or the right (→). In this case we still have to get a result which is a half-integer times Planck's constant. It will be a 50–50 change of being left or right that 'becomes' one or the other when a measurement is made.

Now comes the real fun. Suppose we do a series of measurements on the same electron. First we start with an electron in a superposition state like the one shown above. We then make a measurement in the vertical direction: we get the answer 'up'. The electron is now in a pure state of spin-up-ness. We then pass it through another measurement, but this time it measures the spin to the left or the right. When we select the electron to be spin-up, it tells us nothing about the horizontal spin. Theory thus predicts a 50–50 outcome of this measurement, as is observed experimentally. Suppose we get that the spin is now to the left. Now our long-suffering electron passes into a third measurement which this time is again in the vertical direction. You might imagine that since we have already measured this component to be in the up direction, it would be in that direction again this time. In fact, this is not the case. The intervening measurement seems to reset the up-down spin; the results of the third measurement are back at square one, with a 50–50 chance of getting up or down.

This is just one example of the kind of irreducible randomness that seems to be inherent in quantum theory. I have discussed it in some detail because I want to come back to some of the deeper ramifications later. But before doing so, I just wanted to discuss briefly the idea that has become synonymous with quantum unpredictability.

Heisenberg's Uncertainty Principle is one of the general rules governing the quantum world and it is one aspect of the theory that has some level of popular recognition. It may also provide evidence of a fundamental flaw in the theory. The Uncertainty principle provides a rule for the maximum amount of knowledge one can have about different aspects of the state of a system. In its usual form it states that the uncertainty in the position of a particle Δx and the uncertainty in its momentum Δp must satisfy the inequality

$$\Delta x \Delta p \geq \frac{h}{4\pi}$$

In essence, the better you know the position of a particle the worse you know its momentum. It is perhaps less well known that uncertainty relations of this type apply to other properties too—in fact, any pair of what are known as conjugate variables. In particular, the uncertainty in energy ΔE and lifetime Δt of a state satisfy a similar relationship. In particle physics there are many short-lived states that do not have a well-defined energy for this reason, and through the famous relationship $E = mc^2$ this means they do not have a well-defined mass. We like to think of particles as being little balls whose masses one can look up in tables, but the world of particle physics is not really like that at all. Worse still, this version of the uncertainty principle means that the classical idea that energy has to be conserved is not true. At least for a short time, one can borrow energy from empty space to create particles that should not exist if classical truths held sway. These are virtual particles, and they are one of the biggest thorns in the side of modern physics. In currently favoured particle theories they clog up empty space to an intolerable degree, producing a vacuum energy disaster which is probably at least as significant as the ultraviolet catastrophe I mentioned at the start of the chapter. In a strange echo of the situation around a hundred years ago, most modern physicists are content to carry on using quantum mechanics despite this glaring inadequacy.

Quantum Probability

So far I have focussed on what happens to single particles when quantum measurements are made. Although there seem to be subtle things going on, it is not really obvious that anything happening is very different from systems in which we simply lack the microscopic information needed to make a definition prediction. But quantum probability does have aspects that do not appear in classical stochastic processes.

At the simplest level, the difference is that quantum mechanics gives us a theory for the wave-function which somehow lies at a more fundamental level of description than classical probabilities. Probabilities can be derived mathematically from the wave-function, but there is more information in ψ than there is in $|\psi|^2$; the wave-function is a complex entity whereas the square of its amplitude is entirely real. If one can construct a system of two particles, for example, the resulting wave-function is obtained by superimposing the wave-functions of the individual particles, and probabilities are then obtained by squaring this joint wave-function. This will not, in general, give the same probability distribution as one would get by adding the one-particle probabilities. For complex entities A and B, $A^2 + B^2 \neq (A + B)^2$. To put this another way, remember that I argued in Chapter 4 that there is only one logically consistent way to construct a consistent probability theory by associating a single real number with any proposition. Quantum theory doesn't have that structure: it involves assigning a complex number (or, equivalently, two real numbers) to each proposition. If one can preserve this complexity then quantum probability does not necessarily give the same kind of results that classical probability does. One can write any complex number in the form $a + ib$ (real part plus imaginary part) or, more usefully, as $re^{i\phi}$, where r is the amplitude and ϕ is called the phase. The amplitude gives the classical probability; quantum probabilities arise whenever the phase information is needed.

To put this other way of putting it yet another way, the 'hypothesis space' which quantum logic deals with is a much more complicated mathematical thing than simple possibility spaces I have discussed so far. Rigourous mathematical treatments require a sophisticated concept known as a Hilbert space. I don't have space to go into a general theory in this level of technical detail, but I hope this makes it relatively clear that the wave-function is somehow more fundamental than the probabilities it generates. Finding situations where the quantum phase of a wave-function is important is not easy. It seems to be quite easy to disturb quantum systems in such a way that the phase information becomes scrambled, so testing the fundamental aspects of quantum theory requires considerable experimental ingenuity. But it has been done, and the results are astonishing.

Let us think about a very simple example of a two-component system: a pair of electrons. All we care about for the purpose of this experiment is the spin of the electrons so let us write the state of this system in terms of states such as $|\uparrow\downarrow\rangle$ which I take to mean that the first particle has spin up and the second one has spin down. Suppose we can create this pair of electrons in a state where we know the total spin is zero. The electrons are indistinguishable from each other so until we make a measurement we do not know which one is spinning up and which one is spinning down. The state of the two-particle system might be this:

$$|\psi\rangle = \frac{1}{\sqrt{2}}|\uparrow\downarrow\rangle - \frac{1}{\sqrt{2}}|\downarrow\uparrow\rangle;$$

squaring this up would give a 50% probability of particle one being up and particle two being down and 50% for the contrary arrangement. This does not look too different from the example I discussed above, but this duplex state exhibits a bizarre phenomenon known as *entanglement*. Suppose we start the system out in this state and then separate the two electrons without disturbing their spin states. Before making a measurement we really cannot say what the spins of the individual particles are: they are in a mixed state that is neither up nor down but a combination of the two possibilities. When they are up, they are up. When they are down, they are down. But when they are only half-way up they are in an entangled state.

If one of them passes through a vertical spin-measuring device we will then know that particle is definitely spin-up or definitely spin-down. Since we know the total spin of the pair is zero, then we can immediately deduce that the other one is spinning in the opposite direction. Passing the other electrons through an identical spin-measuring gadget gives a result consistent with this reasoning. So far there is nothing so strange in this.

The problem with entanglement lies in understanding what happens in reality when a measurement is done. Suppose we have two observers, Dick and Harry, each equipped with a device that can measure the spin of an electron in any direction they choose. Particle 1 emerges from the source and travels towards Dick whereas particle 2 travels in Harry's direction. Before any measurement the system is in an entangled superposition state. Suppose Dick decides to measure the spin of electron 1 in the z-direction and finds it spinning up. Immediately the wave-function for electron 2 collapses into the down direction. If Dick had instead decided to measure spin in the left-right direction and found it 'left' similar collapse would have occurred for particle 2, but this time putting it in the 'right' direction. Whatever Dick does, the result of any corresponding measurement made by Harry has a definite outcome, the opposite to Dick's result. So Dick's decision whether to make a measurement up-down or left-right instantaneously transmits itself to Harry who will find a consistent answer, if he makes the same measurement as Dick.

If, on the other hand, Dick makes an up-down measurement but Harry measures left-right then Dick's answer has no effect on Harry, who has a 50% chance of getting 'left' and 50% chance of getting right. The point is that whatever Dick decides to do, it has an immediate effect on the wave-function at Harry's position; the collapse of the wave-function induced by Dick immediately collapses the state measured by Harry. How can particle 1 and particle 2 communicate in this way?

This riddle is the core of a thought experiment by Einstein, Podolsky and Rosen in 1935 which has deep implications for the nature of the information that is supplied by quantum mechanics. The EPR paradox is that each of the two particles—even if they are separated by huge distances—seems to know exactly what the other one is doing. Einstein called this 'spooky action at a distance' and went on to point out that this type of thing simply could not happen in the usual calculus of random variables. His argument was later tightened considerably by John Bell in a form now known as Bell's theorem.

To see how Bell's theorem works, consider the following informal description. Suppose we have two suspects in prison, say Dick and Harry, who are taken apart to separate cells for individual questioning. We can allow them to use notes, electronic organizers, tablets of stone or anything to help them remember any agreed strategy they have concocted, but they are not allowed to communicate with each other. Each question they are asked has only two possible answers: 'yes' or 'no' and there are only three possible questions, called X, Y, and Z. We can assume the questions are asked independently and in a random order to the two suspects. If the interrogators ask the same question then Dick and Harry give the same answer, but when the question is different they give the same answer 25% of the time. What can the interrogators conclude?

The answer is that Dick and Harry must be cheating: either they have seen the question list ahead of time or are able to communicate with each other without the interrogator's knowledge. If they always give the same answer when asked the same question, they must have agreed on answers to all three questions in advance. But if they are asked different questions then at least two of the three prepared answers—and possibly all of them—must be the same. The probability that they would give the same answer to different questions must then be at least one-third and could be larger; this is a simple example of a Bell inequality. They can only keep the number of such false agreements down if they manage to avoid the rules of classical probability.

This example is directly analogous to the behaviour of the entangled quantum state described above under repeated interrogations about its spin in three different directions. The result of each measurement can only be either 'yes' or 'no'. Each individual answer (for each particle) is equally probable in this case; the same question always produces the same answer for both particles, but the probability of agreement for two different questions is indeed 1/4 and not larger as would be expected if the answers were random. For example one could ask particle 1 'are you spinning up' and particle 2 'are you spinning to the right'? The probability of both producing an answer 'yes' is 25% according to quantum theory. Probably the most famous experiment of this type was done in the 1980s, by Alain Aspect and collaborators, involving entangled pairs of polarized photons, rather than electrons, because these are easier to prepare.

What does it all Mean?

The implications of quantum entanglement greatly troubled Einstein long before the EPR paradox. Indeed the interpretation of single-particle quantum measurement (which has no entanglement) was already troublesome. Just exactly how does the wave-function relate to the particle? What can one really say about the state of the particle before a measurement is made? What really happens when a wave-function collapses? These questions take us into philosophical territory that I have set foot in already; the difficult relationship between epistemological and ontological uses of probability theory.

Thanks largely to the influence of Bohr, in the relatively early stages of quantum theory a standard approach to this question was adopted. In what became known as the *Copenhagen* interpretation of quantum mechanics, the collapse of the wave-function as a result of measurement represents a change in the physical state of the system. Before the measurement, an electron really is neither spinning up nor spinning down but in a kind of quantum purgatory. After a measurement it is released from limbo and becomes definitely something. What collapses the wave-function is something unspecified to do with the interaction of the particle with the measuring apparatus or, in some extreme versions of this doctrine, the intervention of human consciousness.

I find it amazing that such a view could have been held so seriously by so many highly intelligent people. Schrödinger hated this concept so much that he invented a thought-experiment of his own to poke fun at it. This is the famous 'Schrödinger's cat' paradox. In a closed box there is a cat. Attached to the box is a device which releases poison into the box when triggered by a quantum-mechanical event, such as radiation produced by the decay of a radioactive substance. One cannot tell from the outside whether the poison has been released or not, so one does not know whether the cat is alive or dead. When one opens the box, one learns the truth. Whether the cat has collapsed or not, the wave-function certainly does. At this point one is effectively making a quantum measurement so the wave-function of the cat is either 'dead' or 'alive' but before opening the box it must be in a superposition state. But do we really think the cat is neither dead nor alive? Is it not certainly one or the other, but that our lack of information prevents us from knowing which? And if this is true for a macroscopic object like a cat, why cannot it be true for a microscopic system like the double-slit experiment with the electron that I described earlier on?

As I learned at a talk by the Nobel prize-winning physicist Tony Leggett—who has been collecting data on this recently—most physicists think Schrödinger's cat is definitely alive or dead before the box is opened, but that the electron does not go through one slit or the other. But where does one draw the line between the microscopic and macroscopic descriptions of reality? If quantum mechanics works for 1 particle, does it work also for 10, 1000 or 10^{23}?

Most modern physicists eschew the Copenhagen interpretation in favour of one or other of two modern interpretations. One involves the concept of *decoherence*, the idea that the phase information that is crucial to the underlying logic of quantum theory can be destroyed by the interaction of a microscopic system with one of larger size.

In effect, this hides the quantum nature of macroscopic systems and allows us to use a more classical description for complicated objects. This certainly happens in practice, but this idea seems to me merely to defer the problem of interpretation rather than solve it. The fact that a large and complex system tends to hide its quantum nature from us does not in itself give us the right to have a different interpretation of the wave-function for big things than we have for small things.

The other trendy way to interpret quantum theory is one for which I have even less time. It is the so-called 'Many Worlds' interpretation. This asserts that our Universe comprizes an ensemble—sometimes called a multiverse—and quantum probabilities are defined over this ensemble. In effect when an electron leaves its source it travels through infinitely many paths in this ensemble of possible worlds, interfering with itself on the way. We live in just one slice of the multiverse so at the end we perceive the electron winding up at just one point on our screen. Part of this is to some extent excusable, because many scientists still believe that one has to have an ensemble in order to have a well-defined probability theory. If one adopts a more sensible interpretation of probability then this is not actually necessary; probability does not have to be interpreted in terms of frequencies. But the many-worlds brigade goes even further than this. They assert that these parallel universes are *real*. What this means is not completely clear, as one can never visit parallel universes other than our own, but never has there been a clearer example of Ed Jaynes' mind projection fallacy.

It seems to me that none of these interpretations is at all satisfactory, and in the gap left by the failure to find a sensible way to understand quantum reality there has grown a pathological industry of pseudo-scientific gobbledegook. Claims that entanglement is consistent with telepathy, that parallel universes are scientific truths, that consciousness is a quantum phenomena abound in the New Age sections of bookshops but have no rational foundation. Physicists may complain about this, but they have only themselves to blame.

But there is one remaining possibility for an interpretation that has been unfairly neglected by quantum theorists despite—or perhaps because of—the fact that it is the closest of all to commonsense. This view that quantum mechanics is just an incomplete theory, and the reason it produces only a probabilistic description is that it does not provide sufficient information to make definite predictions. This line of reasoning has a distinguished pedigree, but fell out of favour after the arrival of Bell's theorem and related issues. Early ideas on this theme revolved around the idea that particles could carry 'hidden variables' whose behaviour we could not predict because our fundamental description is inadequate. In other words two apparently identical electrons are not really identical; something we cannot directly measure marks them apart. If this works then we can simply use only probability theory to deal with inferences made on the basis of our inadequate information. After Bell's work, however, it became clear that these hidden variables must possess a very peculiar property if they are to describe our quantum world. The property of entanglement requires the hidden variables to be non-local. In other words, two electrons must be able to communicate their values faster than the speed of light. Putting this conclusion together with relativity leads one to deduce that the chain of cause and effect must break down: hidden variables are therefore *acausal*. This is such an unpalatable idea that it seems to many physicists to be even worse than the alternatives, but to me it seems entirely plausible that the causal structure of space-time must break down at some level. On the other hand, not all 'incomplete' interpretations of quantum theory involve hidden variables.

One can think of this category of interpretation as involving an epistemological view of quantum mechanics. The probabilistic nature of the theory has, in some sense, a subjective origin. It represents deficiencies in our state of knowledge. The alternative Copenhagen and Many-Worlds views I discussed above differ greatly from each other, but each is characterized by the mistaken desire to put quantum probability in the realm of ontology.

With the gradual re-emergence of Bayesian approaches in other branches of physics a number of important steps have been taken towards the construction of a truly inductive interpretation of quantum mechanics. This programme sets out to understand quantum probability in terms of the 'degree of belief' that characterizes Bayesian probabilities. Recently, Christopher Fuchs and collaborators have shown that, contrary to popular myth, quantum probability can indeed be understood in this way and, moreover, that a theory in which quantum states are states of knowledge rather than states of reality is complete and well-defined. I am not claiming that this argument is settled, but this approach seems to me by far the most compelling, and it is a pity more people are not following it up. More than anything, this is a matter of scientific methodology. If you insist dogmatically that quantum mechanics is the most complete theory possible you will never find an alternative.

The idea that quantum mechanics might be incomplete does not seem to me to be all that radical. Although it has been very successful, there are sufficiently many problems of interpretation associated with it that perhaps it will eventually be replaced by something more fundamental, or at least different. Surprisingly, this is a somewhat heretical view among physicists: most, including several Nobel laureates, seem to think that quantum theory is unquestionably the most complete description of nature we will ever obtain. That may be true, of course. But if we never look any deeper we will never know.

Beyond Quantum Theory?

Quantum theory underpins the way modern physicists describe matter on all scales from the microscopic world of subatomic particles to the grandest scales of the Universe as a whole. Over the course of the twentieth century, quantum mechanics evolved into a more complete version of itself called quantum field theory. This allows a description to be made not just of particles but of the forces between them. In essence, fermionic particles interact with each other by the exchange of so-called 'gauge' bosons that act as force carriers. The first step was to incorporate the interaction between charged particles by the exchange of virtual photons. Subsequently the electromagnetic and weak nuclear forces were unified by the addition of three further force carriers called the W^+, W^- and Z^0, all of which have now been detected in accelerator experiments. Meanwhile the strong force which holds nuclear particles together was also succumbing to the quantum field approach. The theory of quantum chromodynamics (QCD) portrays this interaction as being mediated by an octet of gauge bosons known as the gluons which bind together fundamental particles known as quarks into heavier states called hadrons, of which the neutron and proton are the most familiar examples. It is hoped that this approach will eventually yield a neat reconciliation of QCD and electro-weak theory in a Grand Unified Theory (GUT).

There are many candidates for such a theory but it is not known yet which, if any, is correct. Differences between the theories only manifest themselves at very high energies and there is insufficient data to decide between the alternatives. The current state of play is a 'standard model' of particle physics which is a rather makeshift structure made by bolting together QCD and electro-weak theory, with a few extra quantum fields thrown in to account for the masses of the gauge bosons; these are the so-called Higgs fields.

Quantum field theory may yet provide an even more compelling framework for the description of particles and interactions, but there is one force that has resisted all attempts to describe it using this language. Gravity is the force of nature that is most familiar to us in our everyday world, but it is completely incompatible with quantum theory. One of the reasons for this is that the mathematical theory of quantum fields relies on a technique known as renormalization to eliminate nasty infinities in the answers. This method does not work with gravity. One consequence of this is that, because of the existence of a teeming background of virtual particles, empty space should be infinitely heavy—the so-called cosmological constant problem. This flaw is to my mind highly reminiscent of the ultraviolet radiation catastrophe and it may be that it leads to similarly revolutionary changes in physics.

If a quantum theory of gravity is ever devised then it may be possible to unify it with the other forces of nature to produce what is sometimes called a Theory of Everything (TOE). Current research in this direction is dominated by the superstring theory, which has some promising aspects but which has not lived up to the hype surrounding it. It has yet to make any firm predictions and is therefore currently untestable. Until it becomes possible to test the theory it remains outside the realm of science. Michael Green, one of the founders of the string theory, recently gave a talk on the subject at a conference at Warwick University. At the end a member of the audience asked how long it would be before the string theory made any predictions. Green shrugged and said 'I don't know. Probably never.' This is an astonishing attitude, which suggests that the string theory seems to have set itself apart from the usual strictures of the scientific method.

Part of the problem is that the string theory is mathematically very difficult and the theory itself is far from unique. Indeed much attention is currently being paid to a thing called M-theory, which is a (hypothetical) master-theory that hypothetically incorporates all the possible hypothetical string theories. The problem is that we can probably only ever devise experiments that can probe the low energy limit of this class of theories. It has recently been estimated that there are about 10^{500} effective theories that emerge from string theory at very low energies. Somewhere among this collection there is bound to be one that describes our Universe. This plethora of theoretical possibilities is sometimes called the string *landscape*. Since it comprizes a huge amount of material that is no use to anyone, among which may be hidden the odd piece of worthwhile stuff, I suggest the word scrap-yard is more descriptive. If this is the best that the string theory can do then it marks the end of the line for Science as we know it.

References and Further Reading

Two fundamental papers on quantum entanglement are:

Aspect, Alain *et al.* (1982). Experimental Test of Bell's Inequalities Using Time-Varying Analyzers, *Physical Review Letters*, 49, 1804–1807.

Bell, John S. (1964). On the Einstein-Podolsky-Rosen Paradox, *Physics* 1, 195–200.

Fascinating and historically important ideas on the interpretation of quantum mechanics are contained in:

Bohm, David. (1984). *Causality and Chance in Modern Physics*. Routledge.

Bohm, David. (1980). *Wholeness and the Implicate Order*, Routledge.

An account of quantum theory steeped in multiversalist doctrine is presented in:

Deutsch, David. (1997). *The Fabric of Reality*, Penguin Books.

More even-handed accounts of the subject are:

Rae, Alastair. (1986). *Quantum Physics: Illusion or Reality?* Cambridge University Press.

Squires, Euan. (1986). *The Mystery of the Quantum World*, Adam Hilger.

First steps to a Bayesian interpretation of quantum probability are outlined in:

Carlton Caves, Christopher F. and Rudiger S. (2002). Quantum Probabilities as Bayesian Probabilities, *Physical Review A*. 65, 1–6.

A beautifully written overview of superstrings that shows with admirable clarity how little there is to back up the theory is presented in:

Greene, Brian. (1999). *The Elegant Universe*, Norton Books.

8

Believing the Big Bang

> Our eyes prefer to suppose
> That a habitable place
> Has a geocentric view,
> That architects enclose
> A quiet Euclidean space:
> Exploded Myths—but who
> Would feel at home astraddle
> An ever expanding saddle?
>
> W.H. Auden, in *After Reading a Child's Guide to Modern Physics*

Cosmology is the most ambitious of all branches of science. It aims to build a coherent unified description of the entire Universe as a single system. This means not just the disposition of everything that exists at a particular time, but also how this current state came about, and how it will evolve into the future. This subject has made tremendous advances in recent times. Remarkable observation developments, such as the Hubble Space Telescope, have revealed the structure of objects so distant that the light we see from them must have set out billions of years ago. These discoveries are rightly applauded in the popular press and broadcast media alike, and the general public seem generally fascinated by them. The ability to see ten billion light-years across the Universe rightly fills us with wonder.

But there is also a worrying side to the media portrayal of current cosmology. It seems to me that for every sensational astronomical image published in the newspapers, there is also a story claiming that someone or other has shown that the entire theoretical basis of modern cosmology must be wrong. There is an abundance of crank books supporting 'alternative' theories of creation, many of them profoundly unscientific, and most selling much better than more orthodox treatments. It has to be said that the proliferation of these dissenting views is encouraged by excessively dogmatic pronouncements made by some bona fide cosmologists who claim they have 'proved' theoretical ideas which are purely speculative. Nevertheless there is behind this whole story a lack of understanding of why cosmologists believe what they do. The mathematical ideas involved in the Big Bang theory are so far from everyday human experience that many respond to them as no more than outlandish works of fiction. Why should anyone believe in the Big Bang?

I am a cosmologist, and I do believe in the Big Bang. I do not believe it in some religious sense, but rather in the sense I described in Chapter 4. Given a variety of competing hypothesis, it is more rational to believe in the one that explains the largest number of experimental facts with the smallest number of parameters. In the Bayesian framework the word 'probability' can be used in place of 'reasonable belief'. I think the Big Bang theory is more probable than any other. This does not make it 'right' or 'true'. It may be that in future some other theory fills some of the gaps in modern cosmology, or improves upon the Big Bang's numerous successes. It may also be that forthcoming observations are incompatible with the Big Bang, so that we will be forced to ditch it even if there is no viable alternative.

What I want to do in this Chapter, therefore, is to describe in some detail how the Big Bang model of creation is constructed, and what is the evidence that favours it over the alternatives. Cosmology is unique among the sciences because of the immense scope of its ambitions, but its vastness does not make it qualitatively different from other branches of science. It merely casts the most important features of scientific progress in a sharper relief.

The Basics of the Big Bang

To begin with, I have to outline the structure of the standard framework for scientific cosmology. Cosmology, in some sense of the word, has been around since the dawn of human existence.

Probably every human civilization has wondered about its place in the Universe, why Nature is the way it is, and whether it all could have been different. Many different modes of thought can be applied to these questions. Painters, musicians and writers celebrate Nature and try to convey its relationship to human life through their art. Theologians discuss the idea of Nature as a manifestation of God. Science represents a relatively recent innovation in human learning. In particular, the modern era of scientific cosmology began less than a century ago with the work of Albert Einstein.

To understand how the Big Bang theory is constructed we have to look a little bit at the mathematical theory that underlies it: Einstein's general theory of relativity, which was completed in 1915. Do not worry if you do not understand the theory—few people do. I just want to give some idea of its complexity so that you can have some idea of the historical process by which it arose. If you really do not like the maths at all, you can skip to the next section.

The basic theoretical structure of modern cosmology consists of a family of mathematical models derived from Einstein's general theory of relativity in the 1920s by Alexander Friedmann and George Lemaître, the two founding fathers of the Big Bang. Given the conceptual difficulty of the underlying physics, it is surprising how simple these models are. In essence they contain three components:

(1) a description of the space-time geometry;
(2) a set of equations describing the action of gravity;
(3) a description of the bulk properties of matter.

The most difficult aspect of the theory to understand is probably the first because we are so used to the idea that space has 'normal' geometrical properties: parallel lines never meet; the sum of the angles of a triangle is 180°, and so on. In Einstein's theory space is *curved* in the presence of matter. Light rays travelling near a massive body like a star get bent away from the path they would follow in empty space. Where gravity is particularly strong, space can become so strongly curved that light can be completely trapped. Such regions are called black holes. The behaviour of light rays in cosmological models is illustrated in the Figure.

Even apart from its conceptual difficulties, Einstein's theory is also mathematically very hard. So even if you try not to think about the

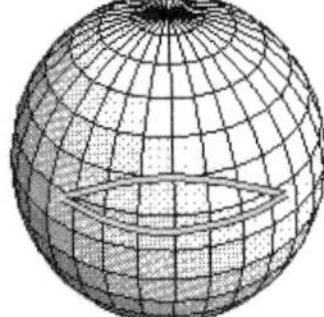

A *closed* universe curves 'back on itself'. Lines that were diverging apart come back together. Density > critical density.

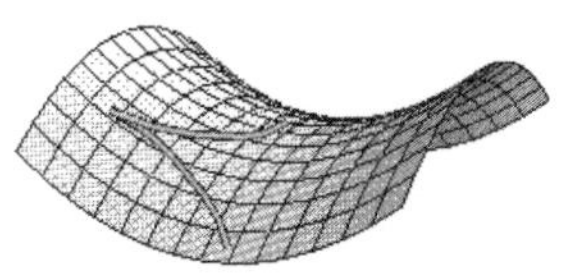

An *open* universe curves 'away from itself'. Diverging lines curve at increasing angles away from each other. Density < critical density.

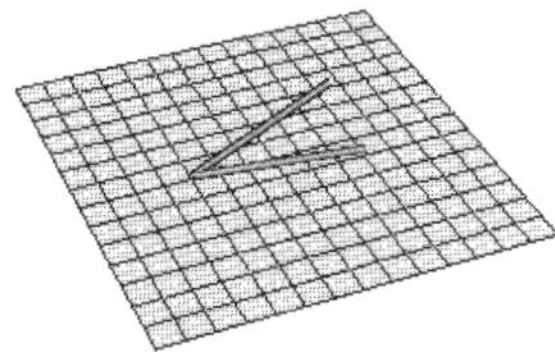

A *flat* universe has no curvature. Diverging lines remain at a constant angle with respect to each other. Density = critical density.

Figure 21 Closed, open and flat universes. There is an intimate connection between the total density of the Universe and its spatial geometry.

concepts, but just turn the handle like many people do with quantum mechanics, you are still likely to make slow progress. Generally speaking, to get anywhere at all with Einstein's theory it is necessary to make some simplifying assumptions about the symmetry of the system to which it is applied. In the case where the system is the entire Universe this need is especially pronounced and the remedy particularly drastic. The *Cosmological Principle* (CP) is the assumption that, on large scales, the Universe is homogeneous and isotropic. The adoption of the CP makes the description of space-time geometry pertaining to cosmological models extremely simple and they can be easily solved.

Einstein's theory involves the use of a metric tensor $g_{\mu\nu}$ that relates four-dimensional space-time intervals to a general set of coordinates in a very flexible way. There is generally no unique to separate space and time. The CP, however, furnishes a preferred time coordinate: any observer can define a 'clock' in terms of the local density of matter at his location. This clock reads what is cosmological time, t. All observers using this clock see the same matter density at any particular time. This simplifies the four-dimensional machinery of Einstein's theory into a much simpler '3 + 1' structure and removes much of the complexity that arises where no special choice of time coordinate is obvious.

Space-times compatible with the Cosmological Principle must have the same local geometry at each point on a surface of constant time. The space-time may be expanding or contracting, however, so different time slices can differ by a scale factor $a(t)$. They can, however, be classified by a single curvature parameter k. There are only three options: $k = 0$ represents a flat universe with a Euclidean geometry on each surface of constant time; $k > 0$ signifies a closed universe, with positively curved spatial surfaces like three-dimensional versions of the surface of a sphere; $k < 0$ indicates negatively curved space sections of hyperbolic form.

Einstein's field equations can be written in the form

$$G_{\mu\nu} = \frac{8\pi G}{c^4} T_{\mu\nu},$$

where the constant G is Newton's constant. This doesn't look too bad at first sight, but the notation is deceptive. The 'tensors' $G_{\mu\nu}$ and $T_{\mu\nu}$ are not individual mathematical quantities but objects with 16 components. The equation above is therefore really 16 equations (actually 10, because not all 16 are independent). Worse still the equations are not at all nice: they are non-linear partial differential equations with unpleasant mathematical properties.

The tensors represent the final two components of the theory I discussed above: $G_{\mu\nu}$ is the Einstein tensor which describes the action of gravity through the curvature of space-time (this contains derivatives of the metric g with respect to space-time coordinates); $T_{\mu\nu}$ is the energy-momentum tensor which describes the bulk properties of matter. These equations are very difficult to deal with, and no general solution exists for arbitrarily complicated distributions of matter and energy.

Applying this general theory to the entire Universe would seem to involve taking on a frightening mathematical problem. However, even a calculation as complicated as this can be simplified by applying a symmetry principle to it. Situations with a high degree of symmetry require fewer variables to describe them. For example, an arbitrary-shaped distribution of matter in three dimensions requires three coordinates. If the distribution is spherically-symmetric, for example in a star, then this can be reduced to only one: the radial distance outward from the centre. The Cosmological Principle imposes a strict symmetry on the Universe, and this drastically simplifies life for theoretical cosmologists. For one thing, the CP forces the matter contents to have the form of a perfect fluid with some pressure p and energy-density ρc^2. For another it means that the four-dimensional space time can be described by one time and one space coordinate. The Einstein equations then simplify to the following set of three ordinary differential equations:

$$3\left(\frac{\dot{a}}{a}\right)^2 = 8\pi G\rho - \frac{3kc^2}{a^2} + \Lambda c^2$$

$$\frac{\ddot{a}}{a} = -\frac{4\pi G}{3}\left(\rho + \frac{3p}{c^2}\right) + \frac{\Lambda c^2}{3}$$

$$\dot{\rho} = -3\left(\frac{\dot{a}}{a}\right)\left(\rho + \frac{p}{c^2}\right).$$

To be precise there are only two independent equations here: the first one can be derived from the other two. The quantity Λ is called the cosmological constant, of which more shortly. These equations determine the time evolution of the cosmic scale factor $a(t)$ which describes the global expansion or contraction of the Universe; the dots denote derivatives with respect to cosmic time. The first of these equations is called the Friedmann equation. If $\Lambda = 0$ it can be derived using Newtonian arguments in which case the curvature constant k is proportional to the total energy of the Universe. Indeed, if $\Lambda \neq 0$ this equation can be obtained by modifying Newton's inverse-square law of gravity via the addition of a term directly proportional to the distance between two objects.

To solve this system, we need to specify the equation of state that characterizes the material contents of the Universe. Cold (non-relativistic matter) can be described by a 'dust' equation of state with $p = 0$. If the Universe is filled with relativistic particles (either photons or very hot matter) then the appropriate equation of state is of the form $p = \rho c^2/3$. In the basic Big Bang theory the early Universe is radiation-dominated; as it expands and cools the matter becomes non-relativistic and the equation of state changes smoothly to that of dust.

So far this probably looks very complicated. How can such a complex system possibly be consistent with the requirement that it be the simplest theory consistent with the data? The mathematics may indeed be difficult, but Einsten's theory is indeed the simplest way of constructing a gravity theory based on the idea of curved space. It does not have adjustable parameters. There are many other variants, such as Brans–Dicke theory, that are similar to general relativity but with additional mathematical functions. Moreover, just think about what this theory incorporates. Because it entails a four-dimensional description of space and time, a solution of the system of equations furnishes a description of the entire evolutionary history of the Universe. It is quite impressive to be able to do that with only two equations!

However, as in the general hypotheses I described in Chapter 4, the Big Bang model does have some free parameters. We have no way of calculating the density of the Universe, or the pressure exerted by its contents, or its global space-time curvature from first principles. These have to be estimated from observations, which I shall describe shortly. The basic framework can describe an entire family of different Universes. Which one do we live in?

Regardless of the specific values of the free parameters in the model, this general framework accounts naturally for Hubble's law, relating the apparent recession velocity v of a galaxy at a distance d, which is of the form $v = H_0 d$. In the Friedmann models this is a consequence of the global expansion of the spatial slices as a function of cosmic time. The Hubble constant H_0 is just $(\dot{a}/a)$ evaluated at the present epoch. This is a rare example of an observational result becoming simpler when viewed through relativistic theory. Hubble's law is actually based on observations of the apparent Doppler shift of galaxy spectra, from which their velocities are inferred. The redshift z is simply the fractional increase in wavelength $(\Delta\lambda/\lambda)$ of a line with wavelength λ measured by an observer at rest with respect to the source of radiation. In cosmology this effect arises as a consequence of the light having travelled along a path through an expanding space-time. If the scale factor increases by a factor $a(t_o)/a(t_s)$ while photons are in transit from a source s to the observer o, they arrive with a wavelength longer than that which they set out with by a factor $1+z = a(t_o)/a(t_s)$. This gives the redshift: red light has longer wavelength than blue light. It is possible now to observe large numbers of quasars with redshifts greater than five or so. Light from these sources set out when the universe was less than one sixth its present size.

Successful though this framework is, it also contains a glaring anomaly. For dust or radiation equations of state, the Universe is always decelerating. Since it is expanding now, it must have been expanding more quickly in the past. There is a stage, at a finite time in the past, at which the scale factor must shrink to zero and the energy density becomes infinite. This is the Big Bang singularity at which the whole framework falls apart. Einstein's equations themselves break down under such extreme conditions. One would expect this (classical) theory to be superseded by a quantum theory of gravity at times earlier than the Planck time, which is of order $(\hbar G/c^5) \approx 10^{-43}$ seconds. Since we do not have such a theory we can say nothing about the Universe's birth. This is why there are free parameters in the theory: we do not know how to set the initial conditions from which the Universe grew.

The Mysterious Vacuum

The cosmological constant Λ was originally introduced by Einstein as a modification of his original version of general relativity. His intention was to provide a remedy for what he regarded as a severe problem with his version of gravity theory. In the absence of the Λ-terms in the cosmic evolution equations, it is not possible to obtain a static solution obeying the CP. For example, one can make the right-hand-side of the Friedmann equation zero by balancing the terms in ρ and k. Having done that, however, one cannot also make the right hand side of the next equation zero. With a Λ-term this difficulty is removed. Einstein thought of this as a modification of the law of gravity (item 2 on the list of requirements for a cosmological model) and he subsequently referred to this as his 'greatest blunder'. His blunder was not so much the introduction of the cosmological constant, but his failure to realize that cosmological models obeying the CP should be dynamical. In other words he failed to predict the Hubble expansion. This error was caused by Einstein's misunderstanding of the available astronomical observations. He thought the Universe could be neither expanding nor contracting because relatively nearby stars were, on average, observed to be neither receding nor approaching the Sun. This is true, but does not contradict the expansion of the Universe. It is only on cosmologically relevant scales, much larger than individual galaxies, where the global expansion wins out over the gravitational effect of local mass concentrations such as galaxies.

Nowadays the cosmological constant is viewed in a very different light. Einstein had modified the left-hand side of his theory, the bit describing gravity and space-time curvature, by the simple addition of a term involving Λ. He could equally well have put this term on the other side of the field equations, in which case it would have been regarded as part of the description of the energy contents. Viewed in this light the cosmological constant (if it really is constant, that is, if it does not evolve with cosmic time) has exactly the same effect as a perfect fluid with an equation of state of the form $p = -\rho c^2$. With the arrival of quantum physics, 'vacuum energy' became the natural way to view the cosmological constant. Following on from the prediction of zero-point vacuum fluctuations now known as the 'Casimir Effect', it was realized that quantum fluctuations in 'empty' space arising from the existence of virtual particles should behave in a very similar way to the classical cosmological constant introduced by Einstein for completely different reasons.

This exciting connection between microscopic quantum physics and large-scale cosmic dynamics is one of the biggest unexplained mysteries in modern science. The problem is that the vacuum energy is formally divergent, so it should utterly dominate the curvature and expansion rate of the Universe. If the divergences are cut off at the Planck energy, which has the enormous value $(hc^5/G)^{1/2} \approx 10^{19}$ GeV, an enormous factor larger than any energy reached in particle experiments, the resulting energy density is 123 orders of magnitude larger than observations allow. This is one of the less impressive predictions of the Big Bang theory, but it signals a gap in the theory rather than a point where the theory is known to be wrong. It seems reasonable to suppose that such a drastic problem has a very fundamental answer. Perhaps there is some reason why the vacuum energy should be exactly zero. Particle physics theories involving supersymmetry suggest this might be possible, owing to a cancellation of fermionic and bosonic contributions. However, as I discuss below, latest cosmological observations suggest this is not correct: there is a non-zero cosmological constant, but it is very small by particle-physics standards.

Thermal History

The evolutionary framework constructed from Einstein's theory has allowed the construction of an astonishingly successful broad-brush description of the evolution of the Universe that accounts for most available observational data. Besides the Hubble expansion, the main evidence in favour of the Big Bang theory was the discovery, by Arno Penzias and Robert Wilson in 1965, of the cosmic microwave background radiation (CMB). Although the spectrum of this radiation was initially not very well determined, in the early 1990s the COBE satellite revealed an astonishingly accurate black-body behaviour with a temperature around 2.7 K. This observation tells us about an important source of pressure which allows us to refine the Big Bang framework.

The importance of the CMB for the advancement of the Big Bang model relates to its physical origin. It is astonishingly difficult to make black-body radiation in a laboratory, especially if the required temperature is as low as 2.7 K. Attempts to reproduce this spectrum using local astronomical sources struggle even harder. One can produce something approaching the correct shape by assuming that light originally produced by stars is scattered on its way to us by clouds of iron whiskers. But this introduces several new parameters, such as the number density and size of the whiskers. The Big Bang accounts for the cosmic radiation background in a much more economical manner than this. Using the Friedmann models we can turn the clock back to an epoch when the Universe was about one-thousandth of its present size at which point the radiation temperature was a thousand times higher. This would be hot enough to ionize atomic matter, which would then have scattered radiation well enough to be effectively opaque like the inside of a star. Although it is very difficult to imagine how such an accurate black-body spectrum could have been made naturally at the low temperature at which it exists now, a dense plasma can maintain sufficient thermal contact with radiation through scattering processes to establish an accurate thermal spectrum. In the Big Bang the cosmic radiation has a black-body form because the Universe behaved like a black body. If it walks like a duck and quacks like a duck, in the absence of any other information, it is not unreasonable to infer that it is a duck.

The cosmic radiation background last interacted with matter about 300,000 years after the Big Bang and has travelled to us freely since then through the expanding space-time. In the meantime, the Universe expanded by about a factor 1000. What we see today as a cold background a few degrees above absolute zero was produced when the whole Universe had a temperature of a few thousand degrees. This is about the same temperature as the surface of a star.

We can turn the clock back even further than this, to an epoch less than a minute after the Big Bang when the temperature of the Universe is measured in billions of degrees. In this brief period the Universe resembles a thermonuclear explosion, during which protons and neutrons are rapidly baked into helium and other light nuclei. In the 1940s George Gamow and others showed that in such conditions one can create helium much more easily than is accomplished by stellar hydrogen burning and with a much lower contamination of heavier elements. The latest observational data suggest that helium accounts for about 24% of the normal matter in the Universe. Most of the rest is hydrogen, and there is just a smattering of other elements like carbon, nitrogen and oxygen. It is not possible to account for this strange chemical composition by saying that all the helium was made in stars during the more recent history of the Universe. If that were the case there would be a lot more other stuff, since stars make heavier elements as well as helium.

The production of the CMB and the synthesis of helium take place during conditions that are achievable in laboratory experiments. To take the Big Bang model further back into the early Universe one needs to adopt more speculative theoretical considerations. For example, when the Universe was around a microsecond old is the time when the standard model of particle physics predicts that heavy particles, such as protons and neutrons, split up into their constituent parts, called quarks and gluons. Earlier still the matter contents of the Universe are expected to be described by a Grand Unified Theory (GUT), and so on, with increasing levels of speculation, back to the Planck time. It must be stressed, however, that these more recent developments do not have the same status as the longer-established ones. Many of them are exciting ideas, and many may well eventually gain strong observational support. But at the moment they are ideas.

One of the most influential ideas of the last 25 years in cosmological theory is called *cosmic inflation*. The first fully-formulated version of this idea was produced by Alan Guth, though some of its components had been advanced by Alex Starobinsky a couple of years earlier and important contributions were also made by Andy Albrecht and Paul Steinhardt. In Guth's idea there is a stage of the evolution of the Universe during which its energy density is dominated not by matter or relativistic particles but by the action of a scalar quantum field, φ. Such a field does not possess the usual equation of state, but is instead characterized by an effective energy density ρc^2 and pressure p which depend on the form of its interaction potential $V(\varphi)$:

$$\rho c^2 = \frac{1}{2}\dot{\varphi}^2 + V(\varphi) \qquad p = \frac{1}{2}\dot{\varphi}^2 - V(\varphi)$$

Guth realized that if a situation could be engineered in which the kinetic terms (involving derivatives of φ) could be made small compared to the potential terms, one could obtain an equation of state of the form $p = -\rho c^2$ just like the vacuum energy described above. If this happens the Universe can be made to accelerate, at least while the field remains in that configuration. This would cause the scale factor to increase exponentially for a very short time driving the curvature term in the Friedmann equation to zero and rendering the spatial surfaces flat to high accuracy, however curved they were before the onset of inflation. After this extravagant but temporary use of the gas pedal, the Universe would be expected to revert to the more sedate radiation-dominated form. This basic concept has led to a plethora of variations: old inflation, new inflation, chaotic inflation, extended inflation, and so on. Among many other things, inflation supplies a resolution of the 'flatness' problem I discuss in the next Chapter.

The realization that our Universe may be accelerating, even during its present old age has resulted in a different manifestation of this idea, that of quintessence. In models based on this idea the cosmological constant (or vacuum energy) is not constant but dynamical and produced by an evolving scalar field at much lower energies than the GUT scales probably involved in inflation.

Inflation and its descendants have an interesting status in cosmological theory. Nobody has yet identified the scalar field responsible, or even whether there is any such thing in our fundamental description of particles and fields. It is also not clear what predictions inflation actually makes. Although I like the idea of inflation, I regard it as part of the penumbra of the Big Bang theory. In other words, if inflation is shown to be wrong—and since it does not make any clear predictions it is hard to see how this could happen soon—it would not invalidate the Big Bang itself. On the other hand, if its shortcomings are remedied then the Big Bang will be closer to a complete theory of the Universe.

Universal Complexity

One of the most important concrete things that inflation has achieved has been to bring us closer to an explanation of how the galaxies and large-scale structure of the Universe came into being and evolved. So far the story has focussed on a broad-brush description of the cosmos in terms of models obeying the Cosmological Principle. In reality, our Universe is not at all smooth and featureless. It contains stars, galaxies, clusters and superclusters of galaxies, and a vast amount of complexity. Where did this structure come from? The basic idea dates back to Sir James Jeans, who we met in Chapter 6. While a perfectly smooth distribution of matter will stay that way forever, any irregularities—however slight—will be amplified in the course of time by the action of gravity. A blob with higher-than-average density will tend to attract material from its surroundings, getting denser still. As it gets denser it attracts more yet more matter. This 'gravitational instability' eventually turns small initial irregularities into highly concentrated clumps held together by their own internal gravitational forces. This process is exponentially fast if the starting configuration is static, so that even microscopically small perturbations grow rapidly to macroscopic size. In an expanding Universe, however, gravity must overcome the expansion in order to initiate collapse. This means that cosmological structure formation is a relatively reluctant process that requires a significant initial 'kick' to get it going.

It also helps if the Universe is dense, since the more matter there is the more gravity has to act upon. It helps even more if the material making up the cosmic density is cold, because then its particles are moving slowly and are easily collected together by gravity. This led theorists to postulate that there must be some form of dark matter which is both abundant and sluggish, like Sunday drivers. This hypothetical 'cold' dark matter (CDM) is still the favoured possibility for most of the stuff in the Universe. It has not yet been detected directly, but its presence is inferred from the gravity that it exerts. As we shall see shortly, however, we do not think there is quite enough of this material to close the Universe but it is still probably the dominant form of matter.

Until the idea of inflation the origin of the initial seed fluctuations was unknown. Nowadays, however, there is a widely-accepted theory that they might have arisen during the Universe's inflationary phase. The hypothetical scalar field does not behave in the smoothly regular fashion that is described by classical physics. Instead it endures fluctuations whose level is governed by the Heisenberg Uncertainty principle, and they arise in the same manner as the zero-point oscillations I discussed in Chapter 7. Of course these phenomena arise on subatomic length scales, but inflation nevertheless renders them visible. The Universe expands by such an enormous factor that a region smaller than a single atom is blown up to a volume as large as our entire Universe.

The extraordinary idea that all the rich complexity of our observable Universe—galaxies, stars, planets and ourselves—emerged as a chance consequence of a quantum fluctuation is both bold and unnerving. Recent observations of the cosmic microwave background have revealed a pattern of temperature fluctuations that seems consistent with this picture. The fluctuations seen are quite small—about one part in a hundred thousand of the cosmic 3° background—but they are precisely of the right amplitude to account for the process of cosmic structure formation. What we see on the microwave sky is a kind of stochastic process and, at least in the simplest inflationary theories, it has Gaussian statistics. As I have remarked previously this is the 'maximum entropy' form for statistical fluctuations with a given variance. In a real sense, our Universe was as disordered as it could be when it was young given that it was so smooth.

That is not to say that we know everything about how the Universe evolved from a nearly featureless expanding ball of gas into a brilliant array of galaxies. While the early stages of structure formation seem to be comprehensible, the complexity involved in the late stages is a real challenge for cosmologists. Heating and cooling of gas, the onset of turbulence, the fragmentation of gas into stars, nuclear ignition, the formation of magnetic fields, and the feedback produced by stellar explosions all combine in a problem of such difficulty that it makes weather forecasting look like a child's puzzle. The processes by which collapsed lumps finally turn into galaxies are so complicated that even the sustained efforts of the most powerful supercomputers have left many questions open. This is a subject for an entire book on its own, so I won't go any further into it here.

Cosmology by Numbers

It will be obvious that the Big Bang 'theory' is seriously incomplete. Because it falls apart at the beginning we have no way of setting the initial conditions that would allow us to obtain a unique solution to the system of equations. There is an infinitely large family of possible universes, so to identify which (if any) is correct we have to use observations rather than pure reason. The simplest way to do this is to re-write the Friedmann equation in a dimensionless form that is a bit easier to swallow than the original:

$$\Omega_{\mathrm{m}} + \Omega_{\Lambda} + \Omega_{k} = 1.$$

The ordinary matter density is expressed via $\Omega_{\mathrm{m}} = 8\pi G\rho/3H^2$, the vacuum energy is $\Omega_{\Lambda} = \Lambda c^2/3H^2$ and $\Omega_k = -kc^2/a^2H^2$. These dimensionless parameters all refer to different components of the energy of the Universe. They can all in principle (and in practice) be measured. Note that Hubble's constant is involved in these definitions so one often measures $\Omega H_0^{\ 2}$. In what follows I will take $H_0 = 100h$ kmsec^{-1} Mpc^{-1} so that h is also dimensionless. In the absence of Λ, Ω_{m} controls the ultimate fate of the Universe as well as its spatial geometry. If $\Omega > 0$ the (closed) Universe eventually recollapses in a 'Big Crunch'. If $\Omega < 0$ it expands forever and is open. Poised between the two is the case $\Omega = 1$, in which the curvature is zero, and the expansion only halts in the infinite future. The behaviour of these different cases is intriguing, and has inspired a great deal of confusion among cosmologists. I will come back to this issue in the next Chapter.

Cosmological nucleosynthesis provides a stringent constraint on the contribution to the matter density arising from 'normal' baryonic material (i.e. anything made up from protons and neutrons). A high baryon abundance would result in more helium than observed, and vice-versa for a low abundance. The value of $\Omega_b h^2$ required to match the observations is 0.012 with a tolerance of about 0.002. In 1994 George Ellis and I wrote a review article for the journal *Nature*, in which we focussed on all the evidence about Ω_m. It is possible to estimate the mean density of the Universe (which is basically what this parameter measures) in many ways, including galaxy and cluster dynamics, galaxy clustering, large-scale galaxy motions, gravitational lensing and so on. At the time many of these 'weighing' techniques were in their infancy but on the basis of what seemed to be the most robust evidence we concluded that the evidence favoured a value of Ω_m between about 0.2 and 0.4. This requires that most of the matter in the Universe be non-baryonic, perhaps in the form of some relic of the early Universe produced at such high energies that it has been impossible to identify in accelerator experiments. As I mentioned above, basic ideas about structure formation theory suggest that this matter should be cold. Our result was somewhat controversial at the time because of the inflationary predilection for flat spatial sections and the (then) general prejudice against Λ. Without a cosmological constant, $\Omega_m = 1$ would be needed to make a flat Universe. As it turns out, the subsequent development of these techniques has not changed our basic conclusion. The last 10 years have, however, seen two stunning observational developments that we did not foresee at all.

In our 1994 paper we devoted only a very small space to the field of 'classical cosmology' which was the mainstay of observational research in this field during its early stages. The idea of this approach is to use observations of distant objects (i.e. sources seen at appreciable redshifts) to directly probe the expansion rate and geometry of the Universe. For example, owing to the focussing effect of spatial curvature, a rod of fixed physical length would subtend a smaller angle when seen through an open Universe than in a closed Universe. A standard light source would likewise appear fainter in a universe that is undergoing accelerated expansion than in a decelerating universe. The difficulty is that standard sources are difficult to come by. The Big Bang universe is an evolving system, so that sources seen at high redshift are probes of an earlier cosmic epoch. Young galaxies are probably very different to mature ones so they cannot be used as standard sources. Classical cosmology consequently fell into disrepute until just a few years ago it underwent a spectacular renaissance.

Two major programmes, *The Supernova Cosmology Project* led by Saul Perlmutter and *The High-z Supernova Search Team* led by Brian Schmidt, exploited the behaviour of a particular kind of exploding star, Type Ia Supernovae, as standard candles. The special thing about this kind of supernova is that it is thought to result from the thermonuclear detonation of a carbon–oxygen white dwarf. These events are themselves roughly standard explosions because the mass involved is always similar, but they are not exactly identical. The breakthrough was the discovery of an empirical correlation between the peak luminosity of the event and the shape of its subsequent light curve. This correlation can be used to reduce the scatter from event to event. The results from both teams seem very conclusive. The measurements indicate that high-redshift supernovae are indeed systematically fainter than one would expect based on extrapolation from similar low-redshift sources using decelerating world models. The results are sensitive to a complicated combination of Ω_{m} and Ω_{Λ} but they strongly favour accelerating world models. This in turn, strongly suggests the presence of a non-zero vacuum energy.

The supernova searches were a fitting prelude to the stunning results that emerged from the Wilkinson Microwave Anisotropy Probe (WMAP) in 2003. The CMB is perhaps the ultimate vehicle for classical cosmology. In looking back to a period when the Universe was only a few hundred thousand years old, one is looking across most of the observable universe. This enormous baseline makes it possible to carry out exquisitely accurate surveying. Although the microwave background is very smooth, the COBE satellite did detect small variations in temperature across the sky. These ripples are caused by acoustic oscillations in the primordial plasma, probably triggered by the primordial quantum processes accompanying inflation. While COBE was only sensitive to long-wavelength waves, WMAP with its much higher resolution, could probe the higher frequency content of the primordial roar. The pattern of fluctuations across the sky, shown in the Figure, encodes detailed information about the modes of oscillation of the cosmic fireball. The spectrum of the temperature variations displays peaks and troughs that contain fantastically detailed data about the basic cosmological parameters described above (and much more). There were strong indications of what WMAP would subsequently find in the earlier experiments, including the balloon-based Maxima and Boomerang, but WMAP was able to map the entire sky rather than small patches. The WMAP results were truly spectacular. Even the preliminary first-year data yield stringent constraints, such as $\Omega_{\mathrm{m}}h^2 = 0.14 \pm 0.02$, $\Omega_b h^2 = 0.024 \pm 0.001$, $h = 0.72 \pm 0.05$ and $\Omega_k = 0.02 \pm 0.02$; these can be

Figure 22 A map of the sky in microwaves revealed by WMAP (the Wilkinson Microwave Anisotropy Probe). According to the prevailing cosmological theory, the pattern of temperature fluctuations contains clues about the quantum origin of galaxies and large-scale structure in the Universe.

strengthened still further by combining the WMAP data with supernovae and galaxy clustering measurements.

These results are consistent with a Universe having flat space, and thus fit naturally within the inflationary paradigm, although there is a strange 30%–70% split in the overall energy budget of the Universe which has yet to be explained by fundamental theory. More importantly from the point of view of this book, the recent WMAP discoveries can be seen to fit neatly in the cycle of theory creation, parameter estimation, and theory evaluation that I outlined in Chapter 4.

The WMAP data have been greeted with a kind of euphoria by cosmologists as the dawn of the era of precision cosmology. The data it has yielded are indeed spectacular, but there are some issues that remain to be resolved. For example, there is a growing realization that the WMAP data contain some strange features, perhaps resulting from some form of foreground contamination. Our own galaxy pollutes the CMB as it itself produces copious dust, synchrotron and free-free emission some of which appears in microwave frequencies. These must be modelled and corrected before one can see the primordial radiation. This is a difficult task, and it remains to be seen how accurately foreground removal can be accomplished in practice with the relatively simple techniques being used to date. Whether the residual foreground is sufficiently important to affect the determination of the basic cosmological parameters seems unlikely, but it may disguise more subtle signatures of exotic early Universe physics. Time will tell.

The strange features that seem to be present in WMAP's initial data release have been presented in the media as evidence that the Big Bang must be wrong. Such misrepresentations are annoyingly premature, since what is currently available is just a preliminary taste of the full set, which will take years to accumulate. It nevertheless remains possible that future data releases will provide evidence that there is an error somewhere in the Big Bang. Contrary to the allegations made by 'alternative' theorists, most mainstream cosmologists would be very excited if this turned out to be the case. We should wait until we understand the data before deciding whether these observations will make the Big Bang more, or less, probable.

The Final Analysis

What I have done so far is to describe the current state of the Universe, or at least of our understanding of it. The interplay between theory and observation over almost a century of dedicated study, has established a 'standard' cosmological model dominated by dark energy and dark matter, with a tiny flavouring of the baryonic matter from which stars, planets and we ourselves are made. This standard model accounts for many precise observations and has been hailed as a spectacular triumph. And so it is.

But this progress should not distract us from the fact that modern cosmology also has a number of serious shortcomings that may take a long time to remedy. I summarize these difficulties here in a series of open questions, most of them fundamental, to which we still do not have answers.

Is General Relativity right? Virtually everything in the standard model depends on the validity of Einstein's general theory. In a sense we already know that the answer to this question is 'no'. At sufficiently high energies (near the Planck scale) we expect classical relativity to be replaced by a quantum theory of gravity. For this reason, a great deal of interest is being directed at cosmological models inspired by the superstring theory. These models require the existence of extra dimensions beyond the four we are used to dealing with. This is not in itself a new idea, as it dates back to the work of Kaluza and Klein in the 1920s, but in older versions of the idea the extra dimensions were assumed to be wrapped up so small as to be invisible. In the latest 'braneworld models', the extra dimensions can be large but we are confined to a four-dimensional subset of them (a 'brane'). In one version of this idea, dubbed the Ekpyrotic Universe, the origin of our observable universe lies in the collision between two branes floating around in a higher-dimensional 'bulk'. Other models are less dramatic, but do result in the modification of the Friedmann equations at early times.

It is not just in the early Universe that departures from general relativity are possible. There remain very few independent tests of the validity of Einstein's theory, particularly in the limit of strong gravitational fields. There is very little independent evidence that the curvature of space time on cosmological scales is related to the energy density of matter. The chain of reasoning leading to the cosmic concordance model depends entirely on this assumption. Throw it away and we have very little to go on.

What is the Dark Energy? The question here is twofold. One part is whether the dark energy is of the form of an evolving scalar field, such as quintessence, or whether it really is constant as in Einstein's original version. This may be answered by planned observational studies such as the SNAP satellite and the United Kingdom's DarkCam, but both of these are at the mercy of government funding decisions. The second part is whether dark energy can be understood in terms of fundamental theory. This is a more open question. At the least it would require an understanding of theory from a particle-physics viewpoint. But it would also require an understanding of why the expected divergence of one-loop fluctuations is suppressed. I think it is safe to say we are still far from such an understanding.

What is the Dark Matter? Around 30% of the mass in the Universe is thought to be in the form of dark matter, but we do not know what form it takes. We do have some information about this, because the nature of the dark matter determines how it tends to clump together under the action of gravity. Current understanding of how galaxies form, by condensing out of the primordial explosion, suggests the dark matter particles should be relatively massive. This means that they should move relatively slowly and can consequently be described as 'cold'. As far as gravity is concerned, one cold particle is much the same as another so there is no prospect for learning about the nature of cold dark matter (CDM) particles through astronomical means unless they decay into radiation or some other identifiable particles. Experimental attempts to detect the dark matter directly are pushing back the limits of technology, but it would have to be a long shot for them to succeed when we have so little idea of what we are looking for.

Did Inflation really happen? The success of concordance cosmology is largely founded on the appearance of 'Doppler peaks' in the CMB fluctuation spectrum. These arise from acoustic oscillations in the primordial plasma that have particular statistical properties consistent with them having been generated by quantum fluctuations in the scalar field driving inflation. This is circumstantial evidence in favour of inflation, but perhaps not strong enough to obtain a conviction.

The smoking gun for inflation is probably the existence of a stochastic gravitational wave background. The identification and extraction of this may be possible using future polarization-sensitive CMB studies even before direct experimental probes of sufficient sensitivity become available. As far as I am concerned, the jury will be out for a considerable time.

Despite these gaps and uncertainties, the ability of the standard framework to account for such a diversity of challenging phenomena provides strong motivation for assigning it a higher probability than its competitors. Part of this motivation is that no other theory has been developed to the point where we know what predictions it can make. There are, for example, many different alternative theories on the market. There are theories based on modifications of Newton's gravitational mechanics, such as MOND, modifications of Einstein's theory, such as the Brans–Dicke theory, theories involving extra dimensions, such as braneworld theory, and so on. Some of these ideas are new, and consequently we do not really understand them well enough to know what they say about observable situations. Others have adjustable parameters so one tends to disfavour them on grounds of Ockham's razor unless and until some observation is made that cannot be explained in the standard framework.

Alternatives should be always explored. The business of cosmology, however, is not only in theory creation but also in theory testing. The great virtue of the standard model is that it allows us to make precise predictions about the behaviour of the Universe and plan observations that can test these predictions. One needs a working hypothesis to target the multi-million-pound investment that is needed to carry out such programmes. By assuming this model we can make rational decisions about how to proceed. Without it we would be wasting taxpayers' money on futile experiments that have very little chance of improving our understanding. Reasoned belief is essential to the advancement of our knowledge. To misquote St. Augustine of Hippo: '*Credo ut intelligam*'; I believe in order that I might understand.

References and Further Reading

For a concise and simplified introduction to modern cosmology, including inflation, see:
Coles, P. (2001). *Cosmology: A Very Short Introduction*, Oxford University Press.

There are very many other introductions to cosmology available, including Jo Silk's *The Big Bang* and Craig Hogan's *The Little Book of the Big Bang*.

For a technical survey of the latest observational and theoretical developments in the field, on which this Chapter is partly based, see:
Coles, P. (2005). The State of the Universe, *Nature*, 433, 248–256.

The review paper about cosmic density referred to in the text is:
Coles, P. and Ellis, G.F.R. (1994) The Case for an Open Universe, *Nature*, 370, 609–615.

Cosmos and its Discontents

> The world is either the result of order or of chance. If the latter, it is cosmos all the same. That is to say it is a regular and beautiful structure.
>
> Marcus Aurelius Antoninus, Mediations, IV. 22

From Universe to universes

The word 'cosmology' is derived from the Greek 'cosmos' which means the world as an orderly system. To the Greeks, the opposite of cosmos was 'chaos'. In their world-view the Universe comprised two competing aspects: the orderly part that was governed by laws and which could be predicted, and the 'random' part which was disordered and unpredictable. To make progress in scientific cosmology we do need to assume that the Universe obeys laws, and that the same laws apply everywhere and for all time. With the rise of quantum theory and its applications to the theory of subatomic particles and their interactions, the cosmology has gradually ceded some of its territory to chaos. In this Chapter I want to explore a few issues relating to the way uncertainty and unpredictability have forced their way into our theories of the Universe. These are the areas where a proper treatment of probability is vital, and why I referred to the dichotomy between cosmos and chaos in the title of this book.

In the early twentieth century, the first systematic world models were constructed based on Einstein's general theory of relativity. This is a classical theory, meaning that it describes a system that evolves smoothly with time. It is also entirely deterministic. Given sufficient information to specify the state of the Universe at a particular epoch, it is possible to calculate with certainty what its state will be at some point in the future. In a sense the entire evolutionary history described by these models is not a succession of events laid out in time, but an entity in itself. Every point along the space-time path of a particle is connected to past and future in an unbreakable chain. If ever the word cosmos applied to anything, this is it.

But as the field of relativistic cosmology matured it was realized that these simple classical models could not be regarded as complete, and consequently that the Universe was unlikely to be as predictable as was first thought. The Big Bang model gradually emerged as the favoured cosmological theory during the middle of the last century, between the 1940s and the 1960s. It was not until the 1960s, with the work of Stephen Hawking and Roger Penrose, that it was realized that expanding world models based on general relativity inevitably involve a break-down of known physics at their very beginning. The so-called singularity theorems demonstrate that in any plausible version of the Big Bang model, all physical parameters describing the Universe (such as its density, pressure, and temperature) become infinite at the instant of the Big Bang. The existence of this 'singularity' means that we do not know what laws if any apply at that instant. The Big Bang contains the seeds of its own destruction as a complete theory of the Universe. Although we might be able to explain how the Universe subsequently evolves, we have no idea how to describe the instant of its birth. This is a major embarrassment. Lacking any knowledge of the laws we do not even have any rational basis to assign probabilities. We are marooned with a theory that lets in water.

The second important development was the rise of quantum theory and its incorporation into the description of the matter and energy contained within the Universe. As I explained in Chapter 7, quantum mechanics (and its development into quantum theory) entails elements of unpredictability. Although we do not know how to interpret this feature of the theory, it seems that any cosmological theory based on quantum theory must include things that cannot be predicted with certainty.

As particle physicists built ever more complete descriptions of the microscopic world using quantum field theory, they also realized that the approaches they had been using for other interactions just would not work for gravity. Mathematically speaking, general relativity and quantum field theory just do not fit together. It might have been hoped that quantum gravity theory would help us plug the gap at the very beginning, but that has not happened yet. What we can say about the origin of the Universe is correspondingly extremely limited and mostly speculative, but some of these speculations have had a powerful impact on the subject.

One thing that has changed radically since the early twentieth century is the possibility that our Universe may actually be part of a much larger collection of Universes. The potential for semantic confusion here is enormous. The Universe is, by definition, everything that exists. Obviously, therefore, there can only be one Universe. In the 'Many Worlds' interpretation of quantum mechanics there is supposed to be a plurality of versions of our Universe, but their ontological status is far from clear. On the other hand, some plausible models based on quantum field theory do admit the possibility that our observable Universe is part of a collection of mini-universes, each of which 'really' exists. This is quite a different thing from the 'quantum ensemble' required by the many worlds doctrine.

According to the Big Bang model, the Universe (or at least the part of it we know about) began about 14 billion years ago. We do not know whether the Universe is finite or infinite, but we do know that if it has only existed for a finite time we can only observe a finite part of it. We cannot possibly see light from further away than fourteen billion light years because any light signal travelling further than this distance would have to have set out before the Universe began. Roughly speaking, this defines our 'horizon': the maximum distance we are able to see. But the fact that we cannot observe anything beyond our horizon does not mean that such remote things do not exist at all. Our observable 'patch' of the Universe might be a tiny part of a colossal structure that extends much further than we can ever hope to see. And this structure might be not at all homogeneous: distant parts of the Universe might be very different from ours, even if our local piece is well described by the Cosmological Principle.

Some astronomers regard this idea as pure metaphysics, but it is motivated by plausible physical theories. The key idea was provided by the theory of cosmic inflation, which I described in the previous Chapter. In the simplest versions of inflation the Universe expands by an enormous factor, perhaps 10^{60}, in a tiny fraction of a second. This may seem ridiculous, but the energy available to drive this expansion is inconceivably large. Given this phenomenal energy reservoir, it is straightforward to show that such a boost is not at all unreasonable. With inflation, our entire observable Universe could thus have grown from a truly microscopic pre-inflationary region. It is sobering to think that everything—galaxy, star, and planet—we can see might be from a seed that was smaller than an atom. But the point I am trying to make is that the idea of inflation opens up one's mind to the idea that the Universe as a whole may be a landscape of unimaginably immense proportions within which our little world may be little more than a pebble. If this is the case then we might plausibly imagine that this landscape varies haphazardly from place to place, producing what may amount to an ensemble of mini-universes. I say 'may' because there is yet no theory that tells us precisely what determines the properties of each hill and valley or the relative probabilities of the different types of terrain.

Many theorists believe that such an ensemble is required if we are to understand how to deal probabilistically with the fundamentally uncertain aspects of modern cosmology. This is not the case. As I tried to explain in Chapter 4, it is perfectly possible to apply probabilistic arguments to unique events like the Big Bang using Bayesian inductive inference. If there is an ensemble, of course, then we can discuss proportions within it, and relate these to probabilities too. Bayesians can use frequencies if they are available, but do not require them. So having, I hope, opened up your mind to the possibility that the Universe may be amenable to a frequentist interpretation, I am going now to explain, with a couple of examples, how one can get along quite nicely without it.

The Flatness Problem

I start with an illustration of how proper discussion of prior probabilities, within a Bayesian framework, can shed important light on fundamental issues connected with the behaviour of the classical Friedmann world models I discussed in the previous chapter. The cosmological 'flatness problem', as it is now known, arises from the peculiar behaviour of these models as they approach the initial singularity and needs to be addressed using careful dynamical arguments. Before doing this, however, I will give an analogy for it which will at least serve to illustrate the qualitative nature of the problem.

Imagine you are standing outside a sealed room. The contents of the room are hidden from you, except for a small window covered by a curtain. You are told that you can open the curtain once and only briefly to take a peep at what is inside, but you may do this whenever you feel the urge. You are told what is in the room. It is bare except for a tightrope suspended across it about two metres in the air. Inside the room is a man who at some time in the indefinite past began walking along the tightrope. His instructions are to carry on walking backwards and forwards along the tightrope until he falls off, either through fatigue or lack of balance. Once he falls he must lie motionless on the floor. You are not told whether he is skilled in tightrope-walking or not, so you have no way of telling whether he can stay on the rope for a long time or a short time. Neither are you told when he started his stint as a stuntman.

What do you expect to see when you open the door? If the man falls off it will take a very short time to drop to the floor. One outcome therefore appears very unlikely: that at the instant you open the curtain, you see him in mid-air between a rope and a hard place. Whether you expect him to be on the rope or on the floor depends on information you do not have. If he is a trained circus artist he might well be capable of walking to and fro along the tightrope for days. If not, he would probably only manage a few steps before crashing to the ground. Either way it remains unlikely that you catch a glimpse of him in gravitational transit.

This probably seems to have very little to do with cosmology, but now forget about tightropes and think about the behaviour of the parameter Ω. To keep things simple, let us ignore the cosmological constant and rearrange things a little so that the Friedmann equation becomes

$$\Omega_{\mathrm{m}} = 1 - \Omega_k.$$

The term on the left-hand side is

$$\Omega_{\mathrm{m}} = \frac{8\pi G\rho}{3H^2},$$

which, in the numerator, says something about the total amount of matter in the Universe and, in the denominator, contains H which measures the cosmic expansion rate. This ratio is usually called the density parameter. It can be written as

$$\Omega_{\mathrm{m}} = \frac{\rho}{\rho_{\mathrm{c}}},$$

where ρ_{c} is called the critical density.

How do we interpret this parameter? Perhaps surprisingly, it is actually quite easy to understand its behaviour using Newtonian concepts. It is all a question of energy. An expanding universe contains huge amounts of positive kinetic energy in the motion of all its constituents away from each other. It also contains gravitational potential energy relating to the gravitational forces pulling them together. The dynamics of the cosmic expansion result from an interplay between these two sources of energy. The total energy of the Universe is fixed. It has to be. Where would it go?

If $\Omega_m < 1$, then $\rho < \rho_c$ and the density of the Universe is insufficient to arrest the initial expansion of the Universe. The universe therefore expands forever: this is called an open universe model. In this case the total energy of the Universe is positive, like an explosion. If, on the other hand $\Omega_m > 1$, then $\rho > \rho_c$ and there is enough matter in the Universe to cause sufficient gravitational pull to stop the initial expansion at some point in the future. The total energy in this case is negative, and this is called a closed universe model. In between, there is the case of $\Omega_m = 1$ where the total energy is exactly zero. This is a flat universe and it has exactly the critical density.

When, in the previous chapter, I discussed the business of estimating Ω_m I implicitly referred to measurements at the current stage of our Universe's history. A key point, however, is that the trade-off between positive and negative energy contributions changes with time. The result of this is that Ω_m is not fixed at the same value forever, but changes with cosmic epoch. As the Universe expands, the matter within it becomes diluted. The expansion rate also changes with time: it slows as the Universe gets older.

Turning these arguments around to consider what happens at the very beginning, it is possible to show that all the Friedmann models begin with Ω_m arbitrarily close to unity at arbitrarily early times, that is, the limit as t tends to zero is $\Omega_m = 1$. In the case in which the Universe emerges from the Big Bang with a value of Ω_m just a tiny bit greater than one then it expands to a maximum at which point the expansion stops, H becomes zero, and the value of Ω_m becomes infinite. Gravitational energy wins out over its kinetic opponent. If, on the other hand, Ω_m sets out slightly less than unity—and I mean slightly, one part in 10^{60} will do—the Universe evolves to a state where Ω_m is very close to zero. In this case kinetic energy is the winner. In the compromise situation with total energy zero, this exact balance always applies. The universe is always described by

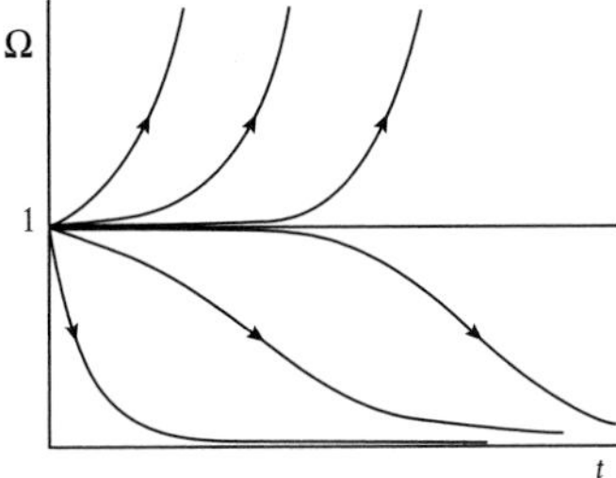

Figure 23 The cosmological flatness problem. If the Universe sets out with Ω exactly equal to unity then it remains that way forever, but any slight disturbance will eventually make it collapse or go into free expansion.

$\Omega_{\mathrm{m}} = 1$. It walks the cosmic tightrope. These cases are illustrated in Figure 23, where the different possible evolutionary paths are shown for slightly different initial conditions.

A slightly different way of describing this is to look at the right-hand side of the simplified Friedman equation. The term $(1 - \Omega_k)$ describes the radius of curvature of the Universe. If the Universe has zero total energy it is flat, so it does not have any curvature at all. If it has positive total energy the curvature radius is finite and positive, in much the same way that a sphere has positive curvature. In the opposite case it has negative curvature, like a saddle.

I hope you can now see how this relates to the curious case of the tightrope walker. If the case $\Omega_{\mathrm{m}} = 1$ applied to our Universe then we can conclude that something trained it to have a fine sense of equilibrium. Without knowing anything about what happened at the initial singularity we might therefore be pre-disposed to assign some degree of probability that this is the case, just as we might be prepared to imagine that our room contained a skilled practitioner of the art of high-level. On the other hand, we might equally suspect that the Universe started off slightly over-dense or slightly under-dense, at which point it should either have re-collapsed by now or have expanded so quickly as to be virtually empty.

A few years ago, Guillaume Evrard tried to put this argument on firmer mathematical grounds by applying a symmetry principle of the type I discussed in Chapter 4 in order to assign a sensible prior probability to Ω_m based on nothing other than the assumption that our Universe is described by a Friedmann model. The result we got was that

$$p(\Omega_m) \propto \frac{1}{\Omega_m(\Omega_m - 1)}$$

I was very pleased with this result, which is based on a principle advanced by Ed Jaynes, but I have no space to go through the mathematics here. Note, however, that this prior has three interesting properties: it is infinite at $\Omega_m = 0$ and $\Omega_m = 1$, and it has a very long 'tail' for very large values of Ω_m. I think of this prior as being the probabilistic equivalent of Mark Twain's description of a horse: 'dangerous at both ends, and uncomfortable in the middle.' It does, however, suggest that we can indeed reasonably assign most of our prior probability to the three special cases I have described.

From recent observations we know, or think we know, that Ω_m is roughly 0.30. To put it another way, this means that the Universe has roughly 30% of the density it would need to have to halt the cosmic expansion at some point in the future. Curiously, this corresponds precisely to the unlikely or 'fine-tuned' case where our Universe is in between the two states in which we might have expected it to lie. I think this may be why so many theoretical cosmologists have, until comparatively recently, resisted the evidence that we live in a relatively low-density Universe. Until the mid-1980s there was a strong ideological preference for models with $\Omega_m = 1$ exactly, not because of the rather simple argument given above but because of the idea of cosmic inflation, which I have already introduced.

Even if you accept my thesis that $\Omega_m = 1$ is a special situation that is in principle possible, it is still the case that it requires the Universe to have been set up with very precisely defined initial conditions. Cosmology can always appeal to special initial conditions to get itself out of trouble, but it is much more satisfactory if properties of our Universe are explained by understanding the physical processes involved rather than by simply saying that 'things are the way they are because they were the way they were.' The latter statement remains true, but it does not enhance our understanding significantly.

The reasoning behind cosmic inflation admits the possibility that, for a very short period in its very early stages, the Universe went through a phase where it was dominated by a third form of energy, vacuum energy. This forces the cosmic expansion to accelerate. This drastically changes the dynamical arguments I gave above. Without inflation the case with $\Omega_m = 1$ is *unstable*: a slight perturbation to the Universe sends it diverging towards a Big Crunch or a Big Freeze. While inflationary dynamics dominate, however, this case has a very different behaviour. It becomes an *attractor* to which all possible universes converge. Whatever the pre-inflationary initial conditions, the Universe will emerge from inflation with Ω_m very close to unity. Inflation trains our Universe to walk the tightrope.

So how can we reconcile inflation with current observations that suggest a low matter density? The key to this question is that what inflation really does is expand the Universe by such a large factor that the curvature radius becomes infinitesimally small. If there is only 'ordinary' matter in the Universe then this requires that the universe have the critical density. However, in Einstein's theory the curvature is zero only if the total energy is zero. If there are other contributions to the global energy budget besides that associated with familiar material then one can have a low value of the matter density as well as zero curvature. The missing link is dark energy, and the independent evidence we now have for it provides a neat resolution of this problem. Or does it? Although spatial curvature does not really care about what form of energy causes it, it is surprising to some extent that the dark matter and dark energy densities are similar. To many minds this unexplained coincidence is a blemish on the face of an otherwise charming structure.

It can be argued that there are initial conditions for non-inflationary models that lead to a Universe like ours. This is true. It is not logically necessary to have inflation in order for the Friedmann models to describe a Universe like the one we live in. On the other hand, it does seem to be a reasonable argument that the set of initial data that is consistent with observations is larger in models with inflation than in those without it. It is rational therefore to say that inflation is more probable to have happened than the alternative. I am not totally convinced by this reasoning myself, because we still do not know how to put a reasonable measure on the space of possibilities existing prior to inflation. This would have to emerge from a theory of quantum gravity which we do not have. Nevertheless, inflation is a truly beautiful idea that provides a framework for understanding the early Universe that is both elegant and compelling. So much so that I almost believe it.

The Anthropic Principle(s)

Once upon a time I was involved in setting up a cosmology conference in Valencia (in Spain). The principal advantage of being among the organizers of such a meeting is that you get to invite yourself to give a talk and to choose the topic. On this particular occasion, I deliberately abused my privilege and put myself on the programme to talk about the 'Anthropic Principle'. I doubt if there is any subject more likely to polarize a scientific audience than this. About half the participants present in the meeting stayed for my talk and entered into a lively debate afterwards. The other half ran screaming from the room.

Roughly speaking, the Anthropic Principle is the name given to a class of ideas arising from the suggestion that there is some connection between the material properties of the Universe as a whole and the presence of human life within it. The name was coined by Brandon Carter in 1974 as a corrective to the 'Copernican Principle' that man does not occupy a special place in the Universe. A naïve application of this latter principle to cosmology might lead us to think that we could have evolved in any of the myriad possible Universes described by the system of Friedmann equations. The Anthropic Principle denies this, although there are different versions that have different logical structures and indeed different degrees of credibility.

It is not really surprising that there is such a controversy about this particular topic, given that so few physicists and astronomers take time to study the logical structure of the subject, and this is the only way to assess the meaning and explanatory value of propositions like the Anthropic Principle. What I want to do here is to unpick this idea and show how it can be understood in terms of Bayesian reasoning.

Suppose we have a model of the Universe M that contains various parameters which can be fixed by some form of observation. Let U be the proposition that these parameters take specific values U_1, U_2, and so on. Anthropic arguments revolve around the existence of life, so let L be the proposition that intelligent life evolves in the Universe. Note that the word 'anthropic' implies specifically human life but many versions of the argument do not necessarily accommodate anything more complicated than a virus. Using Bayes' theorem we can write

$$P(U \mid L \cap M) = \frac{P(U \mid M)P(L \mid U \cap M)}{\sum_{j'} P(U_j \mid M)P(L \mid U_j \cap M)}$$

I have used U_j to denote any alternative set of parameters to U in order to show that it is necessary to include all these in the normalization of the posterior probability. The dependence of the posterior probability $P(U \mid L \cap M)$ on the likelihood $P(L \mid U \cap M)$ demonstrates that the values of U for which $P(L \mid U \cap M)$ is larger correspond to larger values of $P(U \mid L \cap M)$. Since life is observed in our Universe the model-parameters which make life more probable must be preferred to those that make it less so. To go any further we need to say something about the likelihood and the prior. Here the complexity and scope of the model make it virtually impossible to apply in detail the symmetry principles I discussed in Chapter 4. On the other hand, it seems reasonable to assume that the prior is broad rather than sharply peaked. If our prior knowledge of which universes are possible were so definite then we would not really be interested in knowing what observations could tell us. If now the likelihood is sharply peaked in U then this will be projected directly into the posterior distribution.

We have to assign the likelihood using our knowledge of how galaxies, stars and planets form, how planets are distributed in orbits around stars, what conditions are needed for life to evolve, and so on. There are certainly many gaps in this knowledge; I discuss some of them in the next Chapter. Nevertheless if any one of the steps in this chain of knowledge requires very finely-tuned parameter choices then we can marginalize over the remaining steps and still end up with a sharp peak in the remaining likelihood and so also in the posterior probability. For example, there are plausible reasons for thinking that intelligent life has to be carbon-based, and therefore evolves on a planet. It is reasonable to infer, therefore, that $P(U \mid L \cap M)$ should prefer some values of U. This means that there is a *correlation* between the propositions U and L. Knowledge of one should, through Bayesian reasoning, enable us to make inferences about the other.

It is very difficult to make this kind of argument quantitative, but I can illustrate how the argument works with a simplified example. Let us suppose that the relevant parameters contained in the set U include such quantities as Newton's gravitational constant G, the charge on the electron e, and the mass of the proton m. These are usually termed fundamental constants. Our argument above indicates that there might be a connection between the existence of life and the value that these constants jointly take. This is completely in accord with the methodological scheme I described in Chapter 4. Moreover, there is no reason why this kind of argument should not be used to find the values of fundamental constants in advance of their measurement. The ordering of experiment and theory is merely a historical accident; the process is cyclical. An illustration of this type of logic is furnished by the case of a plant whose seeds germinate only after prolonged rain. A newly-germinated (and intelligent) specimen could either observe dampness in the soil directly, or infer it using its own knowledge coupled with the observation of its own germination. This type, used properly, can be predictive and explanatory.

Let us now consider an example, by Robert Dicke, of this type of anthropic reasoning which is of great historical importance. Using the values of G and m together with Planck's constant (h) and the speed of light (c) we can construct a number that measures the strength of the gravitational interaction between two protons

$$\beta = \frac{Gm^2}{\hbar c} \approx 0.6 \times 10^{-38}.$$

This is a dimensionless number, so its value does not depend on the system of units used. Numbers like this are extremely important in cosmology as their numerical values are really fundamental, while those that depend on the choice of particular man-made units tell us more about the people that designed the system of weights and measures than the Universe itself.

Let us suppose that the Universe has existed for a time t since the Big Bang. The size of the observable Universe is then just the distance that light can have travelled in that time, that is, ct. If the density of the matter in the Universe is ρ and we assume, for the sake of argument, that it is all in the form of protons then the number of protons in the visible part of the Universe is

$$N = \frac{4\pi}{3}\rho\frac{(ct)^3}{m}.$$

For the universe described by the Friedman equation the density varies with time roughly according to

$$\rho \approx \frac{1}{Gt^2},$$

which gives

$$N \approx \frac{c^3 t}{Gm} \approx 10^{80},$$

give or take a factor of a few. This produces a curious coincidence. The number of protons in the Universe seems to be given, roughly, by $1/\beta^2$. The eminent British physicist Paul Dirac constructed an entire theory of fundamental physics based on this and related results. His logic was that in the standard theory of the time there seemed to be no necessary connection between β and N. The measured coincidence in these parameter values should therefore have a very low likelihood and consequently the standard theory should be assigned a very low probability.

But Dirac was barking up the wrong tree. The heavier elements found in our bodies (such as iron, required for existence of haemoglobin) are only created in supernova explosions. The time taken for a supernova to form, t_s, can be estimated using the theoretical lifetime of a typical star, which we estimate using the fact that stellar luminosity is caused by burning nuclear fuel. This gives

$$t_s \approx \left(\frac{\hbar c}{Gm^2}\right)\left(\frac{\hbar}{mc^2}\right),$$

again give or take a factor of a few. This gives $t_s \approx 10^{10}$ years, roughly the same as the current age of the Universe. Since we cannot possibly have evolved earlier in the Universe's history than this we can impose this value for cosmic time in the original equation. We then get

$$N \approx \frac{c^3}{Gm}\left(\frac{\hbar c}{Gm^2}\right)\frac{h}{mc^2} = \left(\frac{Gm^2}{\hbar c}\right)^{-2} = \frac{1}{\beta^2}$$

The correlation between these two seemingly distinct parameters arises because they both depend on cosmic time, and that is selected by anthropic considerations.

This argument is just one example of a number of its type, and it has clear (but limited) explanatory power. Indeed it represents a fruitful application of Bayesian reasoning. The question is how surprised we should be that the constants of nature are observed to have their particular values? That clearly requires a probability based answer. The smaller the probability of a specific joint set of values (given our prior knowledge) the more surprised we should be to find them. But this surprise should be bounded in some way: the values have to lie somewhere in the space of possibilities. Our argument has not explained why life exists or even why the parameters take their values, but it has elucidated the connection between two propositions. In doing so it has reduced the number of unexplained phenomena from two to one. But it still takes our existence as a starting point rather than trying to explain it from first principles.

Arguments of this type have been called *Weak Anthropic Principle* by Brandon Carter and I do not believe there is any reason at all for them to be controversial. They are simply Bayesian arguments that treat the existence of life as an observation about the Universe that is incorporated in Bayes' theorem in the same way as all other relevant data and whatever other conditioning information we have. If more scientists knew about the inductive nature of their subject, then this type of logic would not have acquired the suspicious status that it currently has.

Not all anthropic reasoning is so defensible. Among the more speculative versions of the basic idea is the *Strong Anthropic Principle*. In its strongest form this asserts that the Universe *must* have those properties which allow life to evolve within it. This elevates the existence of life to a law of Nature and requires that only the logically conceivable universes should contain life. This sometimes reveals itself as a teleological argument indicating purpose or design: the Universe exists in order that life should arise within it. It is clear that this type of argument has very a different logical status from the previous ones, which were based on inductive reasoning. The Strong Anthropic Principle is itself a *proposition*, something to which we need to assign a probability. The problem is that it is impossible to do this. The existence of life is certainly compatible with this proposition, but we have no basis on which to argue that the existence of life makes it more or less probable. It is simply not testable. The reason for this fundamental difficulty is that two observations which separately imply the same conclusion need not imply each other. During daylight hours both rain and the absence of direct sunlight imply the existence of cloud, but the absence of direct sunlight does not imply rain.

One might as well see the Strong Anthropic Principle in the same light as the existence of God. The axiom that $P(L \mid G) = 1$ would then lead one to reason for the existence of God on the basis of cosmological observations using Bayes' theorem. I do not think this type of thing advances either science or theology.

On the other hand, there are versions of the Strong Anthropic Principle that may turn out to be usefully predictive. This type of argument is mostly applied to the problem of why the fundamental constants of microscopic physics take the values that they do. We do not know how to calculate from fundamental theory what the masses of elementary particles should be, nor do we know how to predict the strengths of their different interactions. If these parameters varied only slightly from the values we know from experiments then atomic structure would be drastically altered, and there would be no chemistry, let alone biology. However, the 'coincidences' that allow life to exist do not necessarily mean that the corresponding parameter values can only be explained by recourse to a design argument.

As I described in Chapter 7, it is a central idea in modern fundamental physics that the parameters controlling physical laws at the relatively low energies accessible in laboratory experiments may have a dynamical origin. There may be a unifying theory that describes interactions in terms of a grand symmetry principle that applies accurately at high energy, but this symmetry might be 'broken' in our low energy world. The 'electroweak' theory that unifies electromagnetism with nuclear interactions provides a good example of this general idea. Undergraduate physics students generally attend entirely separate lecture courses on the theory of electromagnetism and nuclear physics. In fact the theories of electromagnetism and the nuclear interactions are basically the same, the main difference being that the interaction between electrically charged particles is via the exchange of a photon (which is massless) and that between nuclear particles is via a massive particle. At high energies this distinction becomes irrelevant and there is just one theory that applies to them both: the electroweak theory. The Universe would have been sufficiently in its early stages for the electroweak theory to apply, but as it expanded and cooled we ended up in a low-energy state where things look more complicated. This is a process of symmetry breaking. The high energy state is symmetric, the low-energy state asymmetric. An example is given by a pencil standing on end. In this state it is rotationally symmetrical; it looks the same from all sides. A random

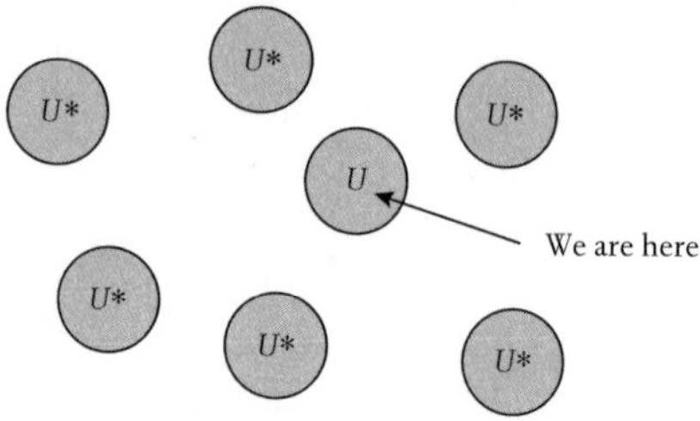

Figure 24 The Strong Anthropic principle may actually be Weak. If the Universe comprises a collection of domains, some of which are habitable (U) and some of which are not then (U*) we must live in one of the habitable set. This means that an unlikely coincidence that might be interpreted in terms of 'design' could be attributable to observer selection.

bump on the table, or unpredictable gust of wind, can topple it into a state where it is lying on the table pointing in a particular direction that does not respect the cylindrical symmetry of the initial configuration.

These ideas can be applied to the Universe as a whole if the idea of symmetry breaking is married to the idea of a spatially varying object such as a quantum field. In this scenario the symmetry breaking in different spatial locations happens independently resulting in different domains where the low energy physics is quite different. In other words the set of fundamental parameters U varies from region to region. Clearly we can only have evolved in a region within which these parameters take values conducive to the existence of life. If this is the case then the Strong Anthropic Principle becomes just a slightly more sophisticated version of the Weak Anthropic Principle. A specific realization of this is Andrei Linde's eternal inflationary universe, in which spontaneous symmetry-breaking is involved so that the 'constants' of nature describing physics below the scale of Grand Unification take different values in mutually incommunicable regions of the Universe. Although it is impossible to visit different 'domains' in order to test this idea experimentally, this does not necessarily render this idea vulnerable to the accusation that it is unscientific. The required dynamical theories might well have other necessary experimental consequences that can be used to test them indirectly.

But it has to be said that all this relies upon physical assumptions about high-energy physics that are yet to be tested in a laboratory. Until we have a fuller understanding of the underlying physical principles, we have to be uncomfortable about the way that we are invoking a multitude of universes to explain the properties of one. If Ockham's Razor applies to parameters then surely it should apply to universes too!

There are other types of argument, such as those based on deeper ideas of fundamental physics. As I explained in Chapter 7, the superstring theory, for example, constrains the number of dimensions of space-time. If we assume it is true then we have to do something about the extra dimensions. This makes no explicit reference to life. Another type of argument is explicitly teleological or causative in the human psychological sense. Probability theory, however, can only deal with correlations between the presence of life and the conditions needed for it. The formulae in physical theories also tell only of correlations between variables. The only valid meaning of causality in physics relates to time-ordering. Much of the confusion surrounding the wackier versions of the Anthropic Principles stems from the failure to distinguish between the human, purposive, meaning of causation and its adoption to mean correlation. Teleological theories typically unite physics and biology. Since these two subjects are complementary and work at different levels of description, theories that accomplish their unification must possess quite extraordinary features. The role of consciousness in collapsing the quantum wave-function is a particularly bizarre example of this. I am therefore highly sceptical about claims that conscious observers can, in some sense, create the Universe they observe. This particular idea has been dubbed the *Participatory Anthropic Principle*. Other versions invoke quantum branching into an ensemble of universe. As I explained in Chapter 7, I regard these as symptoms of frequentist's disease.

More generally one might ask why we assume that the existence of life in the Universe is so important that we enforce it *ab initio*. Much of what I have read about this topic takes it for granted that our existence is somehow so significant that some divine hand must be invoked to explain it. But is life really so important, on a cosmic scale? The presence of Life in the Universe is a relatively recent development. As far as we know, it might be confined to a few isolated outbreaks in damp backwaters of the cosmos, such as Earth. It seems quite likely that its presence in this Universe is temporary, collectively as well as individually. Perhaps we should think of life as a short-term infection contaminating the cosmos during its adolescence but clearing up as it matures. We only think of life as important because we ourselves suffer from it.

This all reminds me of a very clever confidence trick I heard about a few years ago. The perpetrator sent out letters to members of the general public containing advance predictions of the results of forthcoming sporting events. Along with these predictions was the invitation to pay big money to receive future forecasts. Imagine receiving such a letter and subsequently finding all the predictions actually came true. Would you be willing to pay for more red hot tips? Many individuals did, and lost all their money.

The way this scam works is that many different versions of the original letter are sent out, each containing different forecasts. Since the number of possible outcomes of a sporting event is finite, it doesn't take too many letters to cover all the conceivable results. Most people get letters containing predictions that turn out to be false. These individuals throw the letters away and forget about the whole thing. All it costs the trickster for these failures is a bit of postage. However, even if the scammer knows nothing about sport—he might be a physicist, for example,—a small subset receive the correct predictions, and immediately deduce that the person sending them either has psychic powers or inside information. If they knew about all the failed predictions they would not have been so impressed.

References and Further Reading

A Bayesian approach to the flatness problem is presented in:

Evrard, G. and Coles, P. (1995). Getting the Measure of the Flatness Problem, *Classical and Quantum Gravity*, 12, L93–L98.

The most comprehensive account of the different types of Anthropic Principle is:

Barrow, John D. and Tipler, Frank J. (1986). *The Anthropic Cosmological Principle*, Oxford University Press.

Particularly important original papers on this subject are:

Dicke, R.H. (1961). Dirac's Cosmology and Mach's Principle, *Nature*, 192, 440–441.

Carter, B. (1974) Large Number Coincidences and the Anthropic Principle in Cosmology, in *Confrontation of Cosmological Theories with Observation ed. Longair*, M.S., Reidel, Dordrecht, pp. 291–294.

Carr, B.J. and Rees, M.J. (1979) The Anthropic Principle and the Structure of the Physical World, *Nature*, 278, 605–612.

A discussion of the Bayesian aspects of the Weak Anthropic Principle is given in:

Garrett, A.J.M. and Coles, P. (1993). Bayesian Inductive Inference and the Anthropic Cosmological Principle, *Comments on Astrophysics*, 17, 23–47.

10

Life, the Universe and Everything

'Life is very strange' said Jeremy. 'Compared with what?' replied the spider.

Norman Moss, *Men who Play God*

Are we Alone?

Our Universe is certainly contrived in such a way as to make life possible within it. But just because it is possible, that does not mean that it is commonplace. Is life all around us, or did it only happen on Earth? It fascinates me that this topic comes up so often in the question sessions that follow the public lectures I give on astronomy and cosmology. Do you think there is life on other worlds? Are there alien civilizations more advanced than our own? Have extra-terrestrials visited Earth? These are typical of the kind of things people ask me when I give talks on the Big Bang theory of the origin of the Universe. It often seems that people are more interested in finding out if there is life elsewhere than in making more serious efforts to sustain life in the fragile environment of our own planet. But there's no doubting the effect that it would have on humanity to have proof that we are not alone in the cosmos. We could then accept that the Universe was not made for our own benefit. Such proof might also help release mankind from the shackles currently placed on it by fundamentalist religions. But whatever the motives for seeking out life on other worlds, this is undoubtedly a subject worthy of serious scientific study.

Our understanding of the origins of terrestrial life still has important gaps. There is still no compelling direct evidence that life has existed elsewhere in the Solar system. Conditions may, for example, have been conducive to life earlier in the history of Mars but whatever did manage to evolve there has not left any unambiguous clues that we have yet found. The burgeoning new field of astrobiology seeks to understand the possible development of life far from Earth, and perhaps in extreme conditions very different from those found on our planet. This is, however, a very new field and it will be a very long time before it becomes fully established as a rigorous scientific discipline with a solid experimental and observational foundation. What I want to do in this discussion is therefore not to answer the question 'Are we alone?' but to give some idea of the methods used to determine if there might be life elsewhere, including the SETI (Search for Extra-Terrestrial Intelligence) industry which aims to detect evidence of advanced civilizations.

The first ever scientific conference on SETI was held in 1961, in Green Bank, West Virginia, the site of a famous radio telescope. A search had just been carried out there for evidence of radio signals from alien intelligences. This conference did not exactly change the world, which is not surprising because only about ten people showed up. It did, however, give rise to one of the most famous equations in modern science: the Drake Equation.

The astronomer Frank Drake was setting up the programme for the inaugural SETI conference and he wanted to summarize, for further discussion, the important factors affecting the chances of detecting radio transmissions from alien worlds. The resulting equation yields a rough guess of the number of civilizations existing in the Milky Way from which we might get a signal. Of course we cannot calculate the answer. The equation's usefulness is that it breaks down the puzzle into steps, rather than providing the solution. The equation has been modified over the years so that there are various versions of it addressing different questions, but its original form was

$$N = R \times f_p \times n_e \times f_l \times f_i \times f_c \times L$$

The symbols in this equation have the following meanings. N is the number of transmitting civilizations in our Galaxy, which is what we want to determine. The first term on the right hand side is R, which is the birth-rate of stars in our Galaxy per year. We know that the Milky Way is about 10 billion years old, and it contains about 100 billion stars. As a very rough stab we could guess that the required birth-rate is therefore about ten stars per year. It seems unlikely that all stars could even in principle be compatible with life existing in their neighbourhood. For example, very big stars burn out very quickly and explode, meaning that there is very little time for life to evolve there in the first place and very little chance of surviving once it has. Next in the equation is f_p, the fraction of these stars having planets, followed by n_e, the typical number of planets one might find. This is followed by f_l, the fraction of all planets on which life in some form does actually evolve. The next term is f_i the fraction of those planets that have intelligent life on them. Finally we have two factors pertaining to civilization: f_c is the fraction of planets inhabited by intelligent beings on which civilizations arise that are capable of interstellar communication and L is the average lifetime of such civilizations.

The Drake equation probably looks a bit scary because it contains a large number of terms, but I hope you can see that it is basically a consequence of the rules for combining probabilities. The idea is that in order to have a transmitting civilization, you must work out the simultaneous occurrence of various properties each of which whittles away at the original probability, like a form of cosmic bertillonage.

To distil things a little further we can simplify the original Drake equation so that it has only four terms

$$N = N_H \times f_l \times f_c \times f_{now}.$$

The first three terms of the original equation have been absorbed into N_H, the number of habitable planets and the last two have become f_{now}, the fraction of civilized planets that happen to be transmitting now, when we are trying to detect them. This is important because many civilizations could have been born, flourished and died out millions of years in the past so we will never be able to communicate with them.

Whichever way you write it, the Drake equation depends on a number of unknown factors. As we saw in Chapter 2, combining factors multiplicatively like this can rapidly lead to very large (or very small) numbers. In this case each factor is very uncertain, so the net result is very poorly determined.

Recent developments in astronomy mean that we at least have something to go on when it comes to N_H, the number of habitable planets. Until the last decade or so the only planets we knew about for sure were in our own Solar System orbiting our own star, the Sun. We did not know about planets around other stars because even if there were there we were not able to detect them. Many astronomers thought planets would turn out to be quite rare, but then again, absence of evidence is not evidence of absence. Observations now seem to support the idea that planets are fairly common, and this also seems to be implied by our improved understanding of how stars form.

Planets around distant stars are difficult to detect directly because they only shine by light reflected from their parent star and are not themselves luminous. They can, however, be detected in a number of very convincing ways some of which resemble the methods used to detect cosmological dark matter, which I will discuss in the next chapter. Strictly speaking, planets do not orbit around stars. The star and the planet both orbit around their common centre of mass. Planets are generally much smaller than stars so this centre of mass lies very close to the centre of the star. Nevertheless the presence of a planet can be inferred through the existence of a wobble in the star's path through the Galaxy. Dozens of extrasolar planets have been discovered using this basic idea. The more massive the planet, and the closer it is to the star the larger is the effect. Interestingly, many of the planets discovered so far are large and closer in than the large ones in our Solar System (Jupiter, Saturn, Uranus, and Neptune). This could be just a selection effect—we can only detect planets with a big wobble so we cannot find any small planets a long way from their star—but if it is not simply explained away like that it could tell us a lot about the processes by which planets are formed.

The birth of a star is thought to be accompanied by the formation of a flattened disk of debris in the form of tiny particles of dust, ice and other celestial rubbish. In time these bits of dirt coagulate and form larger and larger bodies, all the way up in scale to the great gas giants like Jupiter. The planets move in the same plane, as argued by Laplace way back in the nineteenth century, because they were born in a disk. Of course, while planets may be common we still do not know for sure whether *habitable* planets are also commonplace. We have no reason to think otherwise, however, so we could reasonably assume that there could be one habitable planet per system of planets. This would give a very large value for N_{H}, perhaps 100 billion or so in our Galaxy.

The remaining terms pose a bit more of a problem. We certainly do not have any rational or reliable way to estimate f_{l}. We only know of one planet with life on it. Even Bayesians cannot do much in the way of meaningful statistical inference in this case because we do not have a sensible model framework within which to work. On the other hand, there is a plausibility argument that suggests f_l may be larger rather than smaller. We think Earth formed as a solid object about 4.5 billion years ago. Carbon-isotope evidence suggests that life in a primitive form had evolved about 3.85 billion years ago, and the fossil record suggests life was abundant 3.5 billion years ago. At least the early stages of evolution happened relatively quickly after the Earth was formed and it is a reasonable inference that life is not especially difficult to get going.

It might be possible therefore that $f_l = 1$, or close to it, which would mean that all habitable planets have life. On the other hand, suppose life has a one-in-a-million chance of arising then this reduces the number of potentially habitable planets with life actually on them to only a millionth of this value.

The factor f_c represents the fraction of inhabited planets on which transmitting civilizations exist at some point. Here we really do not have much to go on at all. But there may be some strength in the converse argument to that of the previous paragraph. The fact that life itself arose 3.85 billion years ago but humans only came on the scene within the last million years suggests that this step may be difficult, and f_c should consequently take a small value.

The last term in the simplified Drake equation, f_{now}, is even more difficult because it involves a discussion of the survivability of civilizations. Part of the problem is that we lack examples on which to base a meaningful discussion. More importantly, however, this seems to be a question that causes even distinguished scientists to take leave of their common sense. I will return to a particularly potty version of this tendency towards the end of this Chapter with a discussion of the so-called Doomsday argument. For present purposes, however, it is worth looking at the numbers for terrestrial life. As I mentioned before, the Milky Way is roughly 10 billion years old. We have only been capable of interstellar communication for about 60 years, largely through stray broadcasts of *I love Lucy*. This is only about one part in 200 million of the lifetime of our Galaxy. If we destroy ourselves in the very near future, either by accident or design, then this is our lifetime L as it appears in the original Drake equation. If this is typical of other civilizations then we would have roughly a one in 200 million chance of detecting them at any particular time. Even if our

Galaxy had nurtured hundreds of millions of civilizations, there would only be a few that would be detectable by us now.

Incidentally, it is worth making the comment that Drake's equation was definitely geared to the detection of civilizations by their radio transmissions. It is quite possible that radio-based telecommunication that results in leakage into space only dominates for a brief stage of technological evolution. Maybe some advanced form of cable transmission is set to take over. This would mean that accidental extraterrestrial communications might last only for a short time compared to the lifetime of a civilization. Many SETI advocates argue that in any case we should not rely on accidents, but embark on a programme of deliberate transmission. Maybe advanced alien civilizations are doing this already . . .

In Drake's original discussion of this question, he came to the conclusion that the first six factors on the right-hand-side of the equation, when multiplied together, give a number about one. This leads to the neat conclusion that $N = L$ (when L is the lifetime of a technological civilization *in years*). I would guess that most astronomers probably doubt the answer is as large as this, but agree that the weakest link in this particular chain of argument is L. Reading the newspapers every day does not make me optimistic that L is large.

The Fermi Paradox

One day in 1950 the physicist Enrico Fermi went to lunch with colleagues from the Los Alamos National Laboratory in New Mexico, USA. Fermi had won the Nobel Prize for physics in 1938 for pioneering work he had done in nuclear physics in his home country of Italy until, dismayed by the rise of Fascism under Mussolini, he moved to the United States of America. In Chicago he had been responsible for the construction of the world's first controlled nuclear reactor (on a disused squash court). Over lunch in 1950, however, the conversation turned away from nuclear physics and towards the possible existence of extraterrestrial intelligence. The other physicists present were speculating that there might be a great abundance of other civilizations. Always one to see the wood when others were content to discuss the trees, Fermi's response was a deceptively simple question: 'So where is everybody?' This question, and the difficulties it poses, eventually became known as the Fermi Paradox.

Let us suppose that we use the Drake equation in much the same way that Drake did, that is with some fairly optimistic choices for the actual numbers we need to guess. Accordingly we infer that advanced civilizations are fairly common in the Galaxy. It does not seem too big a step from being able to communicate across interstellar space to being able to build a rocket and actually go there, so let us suppose there may be a sizeable number of civilizations capable of interstellar travel. We also need to imagine that at least some of these civilizations would be interested in exploring and colonizing the galaxy. The essence of Fermi's argument is that, if all this is true, then some of these alien colonists should have reached us by now. I will elaborate on the argument shortly, but it is worth pointing out at the start why it is regarded as a paradox.

It is central to much astronomical thinking that our planet is not in any way special when compared with the hundreds of billions of others potentially cluttering up the Galaxy. If our planet is not special then it makes sense to accept that we are not special either, nor is our civilization. On the other hand, if space colonization is indeed a fairly routine step for advanced societies to take then we should be surrounded by evidence that others have been here already.

Some of you will not be convinced that this is a problem worth thinking about. Others will accept that there is a paradox but will probably be already thinking of ways of escaping it. One obvious solution is to say that the Galaxy has not been colonized because we are the only ones here (and we have not colonized anything yet). If this is true then SETI is a waste of time and money. On the other hand, we have not been around as an advanced civilization very long at all, and even we have managed to send space probes to the Moon and to our neighbouring planets. The nearest star from the Sun is a few light years, and the Galaxy itself is a few tens of thousands of light years across. Even using rockets that travel at a tiny fraction of the speed of light, we could in principle colonize the Galaxy in, say, 10 million years. This is much less than the age of the Milky Way itself.

Even if we could not get our act together sufficiently to send humans on interstellar voyages, we could still perform a kind of colonization using self-replicating robotic probes of a type first suggested by John Von Neumann. We could imagine first sending a few of these machines to the nearest stars. Once there, they would replicate then fan out to spread across the cosmos. Using this technique, any galactic civilization could send out advertisements of its existence very cheaply. One particular physicist, Frank Tipler, has convinced himself using this argument that we are indeed alone. But are there other ways out of the conundrum? Before going through the possibilities we need to look at the argument itself in a bit more detail.

The first essential component of Fermi's argument is that if civilizations are indeed common then many should have arisen before us. I have always assumed that this is an obvious inference, but experience of giving talks on this subject has convinced me that many people do not find it is so, so I will expand on it a little. The key point is one of timescales. We think our Universe began about 14 billion years ago, but there is some uncertainty in this number. To make the numbers nice and simple I will assume that it is actually 12 billion years. This means that the entire span of cosmic history can be mapped neatly onto a normal calendar, with one month representing one billion years.

The Big Bang was on 1 January and we take the present day to be 31 December. Each day on our calendar is about 33 million years of real cosmic time, each hour nearly 1.5 million years, each minute is 23,000 years and each second about 390 years.

According to our current understanding, the Milky Way formed about two billion years into the story, about the same time it takes for something interesting to happen in a novel by Margaret Drabble. This corresponds to mid-February on our calendar. The Earth probably formed about August and by early September it seems primitive life was flourishing. The development of life seems to have reached a kind of crescendo around mid-December with what is called the Cambrian explosion. The plot is not without its twists and turns, however. The dinosaurs appeared around Christmas Day only to disappear forever in the early hours of the morning on 30 December. This was 65 million years ago in real time, but less than two days ago on the calendar.

Most impressive of all is the place of human civilization in this story. The Great Pyramid at Giza in Egypt was built about 13 s ago. Newton's *Principia* was written during the last second. On this timescale I was born about a tenth of a second in the past.

Given this chronology we can look at when we might expect galactic civilizations to develop, if there are any. Assuming that we need Earth-like planets to provide homes for such societies, we first need to be sure that the right materials are available. The early universe was very simple: it was so hot that anything complicated could not exist there. In the first few seconds of the Big Bang—and I mean real time, not cosmic calendar time—the lightest elements were formed. Hydrogen, Helium and small doses of Lithium and Beryllium were synthesized as the Universe expanded and cooled. But the Earth contains larger quantities of much heavier and more complex elements: Carbon, Oxygen, Silicon, Iron, Nickel, even as far as Uranium. These heavier elements are made in stars, and the heaviest of all are only made in stellar explosions called supernovae. In order to supply the raw materials from which the Earth was made, there needed to have been stars around before the Sun was born. Our planet was made by recycling material from a dead star. We do not know exactly how long it might have taken to create the heavy elements we need to make terrestrial planets but stars can be born, die and be recycled on a timescale of a million years or so if they are very massive. This is only a fraction of a day on our calendar so this is not a bottleneck.

The next step is to guess how long it might take a civilization to develop. It has taken about 4.5 billion years for humanity to arise on Earth, so it is not ridiculous to think that the first intelligent civilizations may have come out on the galactic scene about 4.5 billion years after the Milky Way formed. This would be about 6 billion years ago, or around midsummer on the calendar.

Now we have to think about how many civilizations might have arisen during the course of our cosmic year. Here I take a wild guess: a million. If the first of these arose 6 billion years ago then the typical rate at which new ones arise is about one every 6000 years or about three every minute on the cosmic calendar. You may not like that choice. The actual number may be higher or lower than a million. However, I find it very difficult to come up with plausible numbers that lead to a typical timescale for the arrival of new civilizations that is much longer than a few minutes on our calendar. Consequently, there should be thousands that predate ours, many of them by millions or even billions of years.

This does not quite finish the argument, because we need to say something about how rapidly an alien civilization could colonize the Galaxy. I gave a very conservative estimate above that this would take about ten million years or so, but this would shrink rapidly if technology were ever developed that could enable travel at a significant fraction of light speed. The Milky Way could be crossed in about 100,000 years at light speed. But even the longer timescale is much shorter than the time that has been available for civilizations to arise. They really should be here by now. What has gone wrong? There are several possible ways out of this riddle.

One of the most interesting (to me, at any rate) concerns the central assumption that other civilizations would actually *want* to colonize interstellar space. We humans certainly seem to be taking the first steps towards doing this. NASA is devoting much of its future budget to planetary travel, and the European Space Agency is going to spend billions on the Aurora programme in order to send a Frenchman to Mars. Throughout history, humankind has had a tendency to migrate and colonize. A mere hundred thousand years ago our ancestors seem to have been confined to Africa. From there they have spread out all over the surface of the Earth and settled in all kinds of different environments. Sometimes this seems to have been just for the hell of it, and sometimes people have left areas of war or deprivation. Whatever the immediate cause, the urge for exploration is definitely a part of the human psyche. Could this mania for colonization be specific to humans, or do we expect all advanced societies to share it? Will we grow out of it or will it remain part of human nature forever? Could it be that civilizations only stand a chance of surviving into maturity if they realize that their existence depends on caring for the environment that nurtures them rather than squandering their natural resources on interplanetary expeditions?

One argument in favour of space colonization comes up again and again in the popular media. The essence is that population pressure or lack of natural resources will, in the not-too-distant future, require a sizeable fraction of Earth's population to relocate. The figures are impressive. The human population in the year 2000 was somewhere around 6 billion, and is doubling every 50 years or so. In 2150, if nothing changes, it will be 48 billion. Only a few centuries of this kind of population explosion would result in every inch of land being covered with people. Clearly something has to be done. Could we not send the excess population to the colonies?

This idea is a complete non-starter. About 100 million people are born every year right now. Suppose we want to keep the Earth's population constant at 6 billion. To do this we would have to transport 100 million people per year to Mars (or wherever our colony is founded). The biggest transportation vehicle we have right now is the space shuttle, which can carry seven astronauts into an Earth orbit at a cost per launch of $200 million. Even this cannot reach the Moon, never mind another planet, but it makes the point. To send a hundred million people into space, even if there were enough shuttles, would cost two thousand times the entire GDP of the United States every year, forever. Not to mention the tons of greenhouse gases pumped into the atmosphere at every launch. This book contains much that is uncertain, but this is one thing that has absolutely zero probability of working.

There are, of course, several other points of weakness in Fermi's argument that one could choose to attack. Suppose we accept that civilizations are commonplace. We should not forget that we have not actually managed to travel beyond our immediate neighbourhood. We have put a man on the Moon, but that is only a few hundred thousand kilometres away. The few probes we have sent into deep space have not yet escaped the Solar System, and are nowhere near the nearest star. Perhaps it will turn out that interstellar travel, even using von Neumann probes, is just too difficult or expensive in terms of resources for any civilization to bother with. Perhaps civilizations are commonplace, but so is their tendency to self-destruct. Recent events have convinced me that if technology advances to the level where a single individual can hold enough destructive power to annihilate a city, or even a country, then at some point such an event will happen. In effect this means that civilization is a self-limiting process. Once it arises it is bound to destroy itself. Maybe intelligent civilizations are 'out there' and maybe they do survive without blowing themselves up, but maybe also they know about us and have decided not to reveal themselves. Perhaps they are studying us, or perhaps they just have us in a kind of quarantine because we are potentially dangerous. This is a nice idea for science fiction, and indeed forms the basis of what is called the 'Prime Directive' in Star Trek. By definition, we have no evidence that it is true!

The final answer is the most obvious and, at the same time, the most profound. It is that we are, indeed, alone in the Universe.

Doomsday Revisited

I have to put my cards on the table at this point and say that I am very pessimistic about the prospects for humankind's survival into the distant future. Unless there are major changes in the way this planet is governed, our planet may become barren and uninhabitable through war or environmental catastrophe. But I do think the future is in our hands, and disaster is, at least in principle, avoidable. In this respect I have to distance myself from a very strange argument that has been circulating among philosophers and physicists for a number of years. I call it the *Doomsday argument*, and as far as I am aware, it was first introduced by the mathematical physicist Brandon Carter and subsequently developed and expanded by the philosopher John Leslie. It also re-appeared in slightly different guise through a paper in the serious scientific journal *Nature* by the eminent physicist Richard Gott. Evidently, for some reason, some serious people take it very seriously indeed.

The Doomsday argument uses the language of probability theory, but it is such a strange argument that I think the best way to explain it is to begin with a more straightforward problem of the same type.

Imagine you are a visitor in an unfamiliar, but very populous, city. For the sake of argument let's assume that it is in China. You know that this city is patrolled by traffic wardens, each of whom carries a number on his uniform. These numbers run consecutively from 1 (smallest) to T (largest) but you do not know what T is, that is how many wardens there are in total. You step out of your hotel and discover traffic warden number 347 sticking a ticket on your car. What is your best estimate of T, the total number of wardens in the city?

I gave a short lunchtime talk about this when I was working at Queen Mary College, in the University of London. Every Friday, over beer and sandwiches, a member of staff or research student would give an informal presentation about his research, or something related to it. I decided to give a talk about bizarre applications of probability in cosmology, and this problem was intended to be my warm-up. I was amazed at the answers I got to this simple question. The majority of the audience denied that one could make any inference at all about T based on a single observation like this, other than that T must be at least 347.

Actually, a single observation like this *can* lead to a useful inference about T, using Bayes' theorem. Suppose we have really no idea at all about T before making our observation; we can then adopt a uniform prior probability. Of course there must be an upper limit on T. There cannot be more traffic wardens than there are people, for example. Although China has a large population, the prior probability of there being, say, a billion traffic wardens in a single city must surely be zero. But let us take the prior to be effectively constant. Suppose the actual number of the warden we observe is t. Now we have to assume that we have an equal chance of coming across any one of the T traffic wardens outside our hotel. Each value of t (from 1 to T) is therefore equally likely. I think this is the reason that my astronomers' lunch audience thought there was no information to be gleaned from an observation of any particular value, that is $t = 347$.

Let us simplify this argument further by allowing two alternative 'models' for the frequency of Chinese traffic wardens. One has $T = 1000$, and the other (just to be silly) has $T = 1{,}000{,}000$. If I find number 347, which of these two alternatives do you think is more likely? Think about the kind of numbers that occupy the range from 1 to T. In the first case, most of the numbers have 3 digits. In the second, most of them have 6. If there were a million traffic wardens in the city, it is quite unlikely you would find a random individual with a number as small as 347. If there were only 1000, then 347 is just a typical number. There are strong grounds for favouring the first model over the second, simply based on the number actually observed. To put it another way, we would be surprised to encounter number 347 if T were actually a million. We would not be surprised if T were 1000.

One can extend this argument to the entire range of possible values of T, and ask a more general question: if I observe traffic warden number t what is the probability I assign to each value of T? The answer is found using Bayes' theorem. The prior, as I assumed above, is uniform. The *likelihood* is the probability of the observation given the model. If I assume a value of T, the probability $P(t \mid T)$ of each value of t (up to and including T) is just $1/T$ (since each of the wardens is equally likely to be encountered). Bayes' theorem can then be used to construct a posterior probability of $P(T \mid t)$. Without going through all the nuts and bolts, I hope you can see that this probability will tail off for large T. Our observation of a (relatively) small value for t should lead us to suspect that T is itself (relatively) small. Indeed it is a reasonable 'best guess' that $T = 2t$. This makes intuitive sense because the observed value of t then lies right in the middle of its range of possibilities.

Before going on, it is worth mentioning one other point about this kind of inference: that it is not at all powerful. Note that the likelihood just varies as $1/T$. That of course means that small values are favoured over large ones. But note that this probability is uniform in logarithmic terms. So although $T = 1000$ is more probable than $T = 1{,}000{,}000$, the range between 1,000 and 10,000 is roughly as likely as the range between 1,000,000 and 10,000,0000, assuming there is no prior information. So although it tells us something, it does not actually tell us very *much*. Just like any probabilistic inference, there is a chance that it is wrong, perhaps very wrong.

What does all this have to do with Doomsday? Instead of traffic wardens, we want to estimate N, the number of humans that will ever be born, Following the same logic as in the example above, I assume that I am a 'randomly' chosen individual drawn from the sequence of all humans to be born, in past, present, and future. For the sake of argument, assume I number n in this sequence. The logic I explained above should lead me to conclude that the total number N is not much larger than my number, n. For the sake of argument, assume that I am the one-billionth human to be born, that is $n = 1{,}000{,}000{,}0000$. There should not be many more than a few billion humans ever to be born. At the rate of current population growth, this means that not many more generations of humans remain to be born. Doomsday is nigh.

Richard Gott's version of this argument is logically similar, but is based on timescales rather than numbers. If whatever thing we are considering begins at some time t_{begin} and ends at a time t_{end} and if we observe it at a 'random' time between these two limits, then our best estimate for its future duration is of order of how long it has lasted up until now. Gott gives the example of Stonehenge[1], which was built about 4000 years ago: we should expect it to last a few thousand years into the future. Since humanity has been around a few hundred thousand years, it is expected to last a few hundred thousand years more. Doomsday is not quite as imminent as previously, but in any case humankind is not expected to survive sufficiently long to colonize the Galaxy.

[1] Stonehenge is a highly dubious subject for this argument anyway. It hasn't really survived 4000 years. It is a ruin, and nobody knows its original form or function.

You may reject this type of argument on the grounds that you do not accept my logic in the case of the traffic wardens. If so, I think you are wrong. I would say that if you accept all the assumptions entering into the Doomsday argument then it is an equally valid example of inductive inference. The real issue is whether it is reasonable to apply this argument at all in this particular case. There are a number of related examples that should lead one to suspect that something fishy is going on.

There are around 60 million British people on this planet, of whom I am one. In contrast there are 3 billion Chinese. If I follow the same kind of logic as in the examples I gave above, I should be very perplexed by the fact that I am not Chinese. The odds are 50 : 1 against me being British, are they not?

Of course, I am not at all surprised by the observation of my non-Chineseness. My upbringing gives me access to a great deal of information about my own ancestry, as well as the geographical and political structure of the planet. This data convinces me that I am not a 'random' member of the human race. My self-knowledge is conditioning information and it leads to such a strong prior knowledge about my status that the weak inference I described above is irrelevant. Even if there were a million million Chinese and only a hundred British, I have no grounds to be surprised at my own nationality given what else I know about how I got to be here.

This kind of conditioning information can be applied to history, as well as geography. Each individual is generated by its parents. Its parents were generated by their parents, and so on. The genetic trail of these reproductive events connects us to our primitive ancestors in a continuous chain. A well-informed alien geneticist could look at my DNA and categorize me as an 'early human'. I simply could not be born later in the story of humankind, even if it does turn out to continue for millennia. Everything about me—my genes, my physiognomy, my outlook, and even the fact that I bothering to spend time discussing this so-called paradox—is contingent on my specific place in human history. Future generations will know so much more about the universe and the risks to their survival that they will not even discuss this simple argument. Perhaps we just happen to be living at the only epoch in human history in which we know enough about the Universe for the Doomsday argument to make some kind of sense, but know too little to resolve it.

To see this in a slightly different light, think again about Gott's timescale argument. The other day I met an old friend from school days. It was a chance encounter, and I had not seen the person for over 25 years. In that time he had married, and when I met him he was accompanied by a baby daughter called Mary. If we were to take Gott's argument seriously, this was a random encounter with an entity (Mary) that had existed for less than a year. Should I infer that this entity should probably only endure another year or so? I think not. Again, bare numerological inference is rendered completely irrelevant by the conditioning information I have. I know something about babies. When I see one I realize that it is an individual at the start of its life, and I assume that it has a good chance of surviving into adulthood. Human civilization is a baby civilization. Like any youngster, it has dangers facing it. But is not doomed by the mere fact that it is young.

John Leslie has developed many different variants of the basic Doomsday argument, and I do not have the time to discuss them all here. There is one particularly bizarre version, however, that I think merits a final word or two because is raises an interesting red herring. It's called the 'Shooting Room'.

Consider the following model for human existence. Souls are called into existence in groups representing each generation. The first generation has ten souls. The next has a hundred, the next after that a thousand, and so on. Each generation is led into a room, at the front of which is a pair of dice. The dice are rolled. If the score is double-six then everyone in the room is shot and it's the end of humanity. If any other score is shown, everyone survives and is led out of the Shooting Room to be replaced by the next generation, which is ten times larger. The dice are rolled again, with the same rules. You find yourself called into existence and are led into the room along with the rest of your generation. What should you think is going to happen?

Leslie's argument is the following. Each generation not only has more members than the previous one, but also contains more souls than have ever existed to that point. For example, the third generation has 1000 souls; the previous two had 10 and 100 respectively, that is 110 altogether. Roughly 90% of all humanity lives in the last generation. Whenever the last generation happens, there bound to be more people in that generation than in all generations up to that point. When you are called into existence you should therefore *expect* to be in the last generation. You should consequently expect that the dice will show double six and the celestial firing squad will take aim. On the other hand, if you think the dice are fair then each throw is independent of the previous one and a throw of double-six should have a probability of just one in thirty-six. On this basis, you should expect to survive. The odds are against the fatal score.

This apparent paradox seems to suggest that it matters a great deal whether the future is predetermined (your presence in the last generation requires the double-six to fall) or 'random' (in which case there is the usual probability of a double-six). Leslie argues that if everything is pre-determined then we are doomed. If there is some indeterminism then we might survive. This is not really a paradox at all, simply an illustration of the fact that assuming different models gives rise to different probability assignments.

While I am on the subject of the Shooting Room, it is worth drawing a parallel with another classic puzzle of probability theory, the St Petersburg Paradox. This is an old chestnut to do with a purported winning strategy for Roulette. It was first proposed by Nicolas Bernoulli but famously discussed at greatest length by Daniel Bernoulli in the pages of *Transactions of the St Petersburg Academy*, hence the name. It works just as well for the case of a simple toss of a coin as for Roulette as in the latter game it involves betting only on red or black rather than on individual numbers.

Imagine you decide to bet such that you win by throwing heads. Your original stake is £1. If you win, the bank pays you at even money (i.e. you get your stake back plus another £1). If you lose, that is get tails, your strategy is to play again but bet double. If you win this time you get £4 back but have bet £2 + £1 = £3 up to that point. If you lose again you bet £8. If you win this time, you get £16 back but have paid in £8 + £4 + £2 + £1 = £15 up to that point. Clearly, if you carry on the strategy of doubling your previous stake each time you lose, when you do eventually win you will be ahead by £1. It is a guaranteed winner. Is it not?

The answer is yes, as long as you can guarantee that the number of losses you will suffer is finite. But in tosses of a fair coin there is no limit to the number of tails you can throw before getting a head. To get the correct probability of winning you have to allow for *all* possibilities. So what is your expected stake to win this £1? The answer is the root of the paradox. The probability that you win straight off is 1/2 (you need to throw a head), and your stake is £1 in this case so the contribution to the expectation is £0.50. The probability that you win on the second go is 1/4 (you must lose the first time and win the second so it is 1/2 times 1/2) and your stake this time is £2 so this contributes the same £0.50 to the expectation. A moment's thought tells you that each throw contributes the same amount, £0.50, to the expected stake. We have to add this up over all possibilities, and there are an infinite number of them. The result of summing them all up is therefore infinite. If you do not believe this just think about how quickly your stake grows after only a few losses: £1, £2, £4, £8, £16, £32, £64, £128, £256, £512, £1024, etc. After only ten losses you are staking over a thousand pounds just to get your pound back. Sure, you can win £1 this way, but you need to expect to stake an infinite amount to guarantee doing so. It is not a very good way to get rich.

The relationship of all this to the Shooting Room is that it is shows it is dangerous to pre-suppose a finite value for a number which in principle could be infinite. If the number of souls that could be called into existence is allowed to be infinite, then any individual has no chance at all of being called into existence in any generation!

Amusing as they are, the thing that makes me most uncomfortable about these Doomsday arguments is that they attempt to determine a probability of an event without any reference to underlying mechanism. For me, a valid argument about Doomsday would have to involve a particular physical cause for the extinction of humanity (e.g. asteroid impact, climate change, nuclear war, etc). Given this physical mechanism one should construct a model within which one can estimate probabilities for the model parameters (such as the rate of occurrence of catastrophic asteroid impacts). Only then can one make a valid inference based on relevant observations and their associated likelihoods. Such calculations may indeed lead to alarming or depressing results. I fear that the greatest risk to our future survival is not from asteroid impact or global warming, but self-destructive violence carried out by humans themselves. Science has no way of being able to predict what atrocities people are capable of so we cannot make a good assessment of the chances. But the absence of any specific mechanism in the versions of the Doomsday argument

I have discussed robs them of any scientific credibility at all. There are better reasons for worrying about the future than mere numerology.

References and Further Reading

A nice introduction to the possibilities and probabilities that there might be life elsewhere is:

Bennett, Jeffrey, Shostak Seth, and Jakosky, B. (2003). *Life in the Universe*, Addison-Wesley.

The Doomsday argument is discussed by:

Leslie, J. (1965). Is the end of the World Nigh? *Philosophical Quarterly*, 40, 65–72.

A slightly different version is contained in:

Gott, J. Richard, (1993). Implications of the Copernican Principle for our Future Prospects, *Nature*, 363, 315–319.

For a wonderfully written, cogently argued and deeply pessimistic outlook on our survival as a species, see:

Rees, Martin J. (2004). *Our Final Century*, Arrow Books.

11

Summing Up

> If a man will begin with certainties, he shall end in doubts; but if he will be content to begin with doubts, he shall end in certainties . . .
>
> Francis Bacon, in *The Advancement of Learning*

Statistics on Trial

I have taken a very circuitous route around the Natural World so far but in this final Chapter, I want to come back to Earth and discuss some broader aspects of the role of probability in everyday life. I have wandered through such esoteric subjects as chaos theory, quantum mechanics, and the anthropic principle, party because they are topics that I have to work with during my working life as a cosmologist, but also because they are 'safe'. What I mean is that, while these subjects may be greeted with mild interest by the person-in-the-street, they are generally thought to be so distant from the mundanity of human existence that they are not perceived as being threatening. This is one of the reasons why so many popular books on cosmology do well. Other branches of science, such as microbiology, are treated with suspicion or even outright hostility because the outputs of their study may have the potential to influence our lives in a harmful way. While the journey may have taken me in strange directions, I hope the perspectives we encountered on the way will assist in developing a deeper understanding of what goes on in our own backyard.

One particular area that I would like to explore is the role of probability in the courtroom. This subject fascinates me, although the level of my knowledge of legal practice is limited to daytime re-runs of *Perry Mason*. The first thing to say about statistics in jurisprudence is that, generally speaking, it is a complete disaster. Since the person in the street understands so little about probability, it seems obvious that the person in the jury box will fare little better.

Even the so-called expert witnesses that are supposed to help the jury understand scientific evidence are not guaranteed to know what they are talking about when it comes to statistics. I will come to the case of Sir Roy Meadow shortly.

It depresses me very deeply that our legal system rests on such flawed foundations, when the great early thinkers on probability (especially Laplace) had such high hopes that their work would lead to a revolution in this field. The rise of science over the last century or so should have made it possible to improve the standards of legal proof by an enormous factor, but instead many of the most important advances (such as DNA fingerprinting, which I discuss below), are either ruled inadmissible in certain courts or so widely misunderstood that they are of dubious value anyway. The legal profession seems so obsessed by its own procedures that it pays little attention to whether these could be improved, to increase the rate at which the guilty are convicted and decrease the rate at which the innocent are convicted.

In this book I have taken on the role of advocate in my own way: arguing for the wider appreciation and application of Bayesian probabilistic reasoning. I don't think that it is reasonable to suppose that we will ever have juries comprised entirely of experts in inductive logic, so it is pointless to say that everything would be better if everyone was fully trained in mathematical statistics. What I would say is that, in this framework, there are aspects of the current legal system that make no sense whatsoever.

For example, take the legal dictum 'innocent until proven guilty'. In a Bayesian framework this means assigning zero prior probability of guilt to the defendant in advance of the trial. No amount of likelihood can possibly defeat a zero prior, so all defendants should be acquitted if this principle is adopted. Assuming that it has been definitely established that a crime has been committed, the probability of a particular person having committed it is better assigned as the reciprocal of the number of people in the population. Subsequent data (such as witness statements, fingerprints, mobile phone records, and so on) may increase or decrease this probability.

The second aspect of legal practice that baffles me is the requirement that, in criminal trials, the defendant must be found guilty 'beyond reasonable doubt'. As a scientist, it is my job to have reasonable doubt about nearly everything so if I were to take this seriously I could never convict anyone of anything. (I have never been on a jury, and would probably get myself into trouble for contempt of court if I were). More seriously, what is 'beyond reasonable doubt'? In Chapter 4, I referred to the use of significance levels in hypothesis testing. Many statisticians seem happy to use 5%, or about 20-1 against, which in some sense represents their degree of doubt. Is this a reasonable level? I have had horses win at 20-1. Would you be happy to convict someone to life imprisonment on the basis that a 20-1 shot has no chance at all? Whatever happens, the verdict of a jury will have some element of doubt attached to it, even if there is a confession: who is to say the defendant was not subjected to unbearable pressure?

Furthermore, the realization that there must be some uncertainty in any verdict furnishes what I believe to be a cast-iron argument against the death penalty, even for the most grisly and extreme murders. One cannot on rational grounds justify an infinite punishment when there is a finite probability that a mistake has been made. If the death penalty had been in force within the United Kingdom during the 1970s and 1980s, the Birmingham Six and Guildford Four would definitely have been executed. We now know that all ten of these individuals were innocent, and that they were convicted because of errors in forensic evidence and confessions fabricated by the police.

In civil court cases, the burden of proof is rather different. In a dispute between two individuals the jury is usually required to find a verdict where the balance of probabilities lies. This seems much more sensible as a general rule. The reason it does not apply to state prosecutions is presumably to prevent frivolous or malicious prosecutions where there is clearly reasonable doubt at the outset. To prevent repeated abuses of this type there is also a 'double-jeopardy' rule which prevents an individual being subject to repeated trials for the same offence after being acquitted even if new evidence is obtained that was not available at the original trial. I understand the need for this type of regulation: for one thing, it puts the onus on the prosecution to assemble the best possible case at the original trial. But it does not allow for the possibility that subsequent scientific discoveries may lead to new methods that could have provided compelling evidence at the time of the trial had they been invented then. It is possible under the British legal system for later evidence to be used in the Court of Appeal to quash the convictions of innocent people or to arrange a re-trial, but it is not possible to try an acquitted defendant again under the same circumstances.

DNA Fingerprinting

I want to turn to a specific example of forensic statistics which has been involved in some high-profile cases and which demonstrates how careful probabilistic reasoning is needed to understand scientific evidence. Typically, the use of DNA evidence involves the comparison of two samples: one from an unknown source (evidence, such as blood or semen, collected at the scene of a crime) and a known or reference sample, such as a blood or saliva sample from a suspect. If the DNA profiles obtained from the two samples are indistinguishable then they are said to 'match' and this evidence can be used in court as indicating that the suspect was the origin of the sample.

In courtroom dramas DNA matches are usually presented as being very definitive. In fact, the strength of the evidence varies very widely depending on the circumstances. If the DNA profile of the suspect or evidence consists of a combination of traits that is very rare in the population at large then the evidence can be very strong that the suspect was the contributor. If the DNA profile is not so rare then it becomes more likely that both samples match simply by chance. This probabilistic aspect makes it very important to understand the logic of the argument very carefully.

So how does it all work? A DNA profile is not a complete map of the entire genetic code contained within the cells of an individual, which would be such an enormous amount of information that it would be impractical to use it in court. Instead, a profile consists of a few (perhaps half-a-dozen) pieces of this information called *alleles*. An allele is one of the possible codings of DNA of the same gene at a given position (or locus) on one of the chromosomes in a cell. A single gene may, for example, determine the colour of the blossom produced by a flower; more often genes act in concert with other genes to determine the physical properties of an organism. The overall physical appearance of an individual organism, that is any of its particular traits, is called the phenotype and it is controlled, at least to some extent, by the set of alleles that the individual possesses. In the simplest cases, however, a single gene controls a given attribute. The gene that controls the colour of a flower will have different versions: one might produce blue flowers, another red, and so on. These different versions of a given gene are called alleles.

Some organisms contain two copies of each gene; these are said to be diploid. These copies can either be both the same, in which case the organism is homozygous, or different in which case it is heterozygous; in the latter case it possesses two different alleles for the same gene. Phenotypes for a given allele may be either dominant or recessive (although not all are characterized in this way). For example, suppose the dominat and recessive alleles are called A and a respectively. If a phenotype is dominant then the presence of one associated allele in the pair is sufficient for the associated trait to be displayed, that is AA, aA and Aa will both show the same phenotype. If it is recessive, both alleles must be of the type associated with that phenotype so only aa will lead to the corresponding traits being visible.

Now we get to the probabilistic aspect of this. Suppose we want to know what the frequency of an allele is in the population, which translates into the probability that it is selected when a random individual is extracted. The argument that is needed is essentially statistical. During reproduction the offspring assemble their alleles from those of their parents. Suppose that the alleles for any given individual are chosen independently. If p is the frequency of the dominant gene and q is the frequency of the recessive one, then we immediately write:

$$p + q = 1.$$

Using the product law and assumed independence, the probability of homozygous dominant pairing (i.e. AA) is p^2, while that of the pairing aa is q^2. The probability of the heterozygotic outcome is $2pq$ (the two possibilities, each of probability pq are Aa and aA). This leads to the result that

$$p^2 + 2pq + q^2 = 1.$$

This called the Hardy-Weinberg law. It can easily be extended to cases where there are two or more alleles, but I won't go through the details here.

Now what we have to do is examine the DNA of a particular individual and see how it compares with what is known about the population. Suppose we take one locus to start with, and the individual turns out to be homozygotic: the two alleles at that locus are the same. In the population at large the frequency of that allele might be, say, 0.6. The probability that this combination arises 'by chance' is therefore 0.6 times 0.6, or 0.36. Now move to the next locus, where the individual profile has two different alleles. The frequency of one is 0.25 and that of the other is 0.75. so the probability of the combination is '$2pq$', which is 0.375. The probability of a match at both these loci is therefore 0.36 times 0.375, or 13.5%. The addition of further loci gradually refines the profile, so the corresponding probability reduces. This is a perfectly *bona fide* statistical argument, provided the assumptions made about population genetics are correct. Let us suppose that a profile of seven loci leads to a probability of one in ten thousand for a particular individual. Now suppose the profile of our suspect matches that of the sample left at the crime scene. This means that, either the suspect left the trace there, or an unlikely coincidence happened: that, by a 1 : 10,000 chance, our suspect just happened to match the evidence.

This kind of result is often quoted in the newspapers as meaning that there is only a 1 in 10,000 chance that someone other than the suspect contributed the sample or, in other words, that the odds against the suspect being innocent are ten thousand to one against. Such statements are gross misrepresentations of the logic, but they have become so commonplace that they have acquired their own name: the 'Prosecutor's Fallacy'.

To see why this is wrong, imagine that whatever crime we are talking about took place in a big city with 1,000,000 inhabitants. How many people in this city would have DNA that matches the profile? Answer 1 in 10,000 of them which comes to 100. Our suspect is one. In the absence of any other information, the odds are roughly 100 : 1 against him being guilty rather than 10,000 : 1 in favour. In realistic cases there will of course be additional evidence that excludes the other 99 potential suspects, so it is incorrect to claim that a DNA match actually provides evidence of innocence. This converse argument has been dubbed the Defence Fallacy, but nevertheless it shows that statements about probability need to be phrased very carefully if they are to be understood properly by lay people.

The Dangers of Medical Statistics

Although modern cosmology requires a great deal of complicated statistical reasoning, I have it relatively easy because there is not much chance that any errors I make will harm anyone. Speculations about the anthropic principle or theories of everything are unlikely to be reported in the mass media. If they are, and are garbled, the resulting confusion is unlikely to be fatal. The same can not be said of the field of medical statistics. I can't resist the opportunity to include an example of how a relatively simple statistical test can lead to total confusion. In this version, it is known as Simpson's Paradox.

A standard thing to do in a medical trial is to take a set of patients suffering from some condition and divide them into two groups. One group is given a treatment (T) and the other group is given a placebo; this latter group is called the control and I will denote it T* (no treatment). To make things specific suppose we have 100 patients, of whom 50 are actively treated and 50 form the control. Suppose that at the end of the trial for the treatment, patients can be classified as recovered ("R") or not recovered ("R*"). Consider the following outcome, displayed in a contingency table:

	R	R*	Total	Recovery
T	20	30	50	40%
T*	16	34	50	32%
Totals	36	64	100	

Clearly the recovery rate for those actively treated (40%) exceeds that for the control group, so the treatment seems to produce some benefit.

Now let us divide the group into older and younger patients: the young group Y contains those under 50 years old (carefully defined so that I would belong to it) and Y* is those over 50.

The following results are obtained for the *young* patients.

	R	R*	Total	Recovery
T	19	21	40	47.5%
T*	5	5	10	50%
Totals	24	26	50	

While the older group returns the following data.

	R	R*	Total	Recovery
T	1	9	10	10%
T*	11	29	40	27.5%
Totals	12	38	50	

For each of the two groups, the recovery rate for the control exceeds that of the treated patients. The placebo works better than the treatment for the young and the old separately, but for the population as a whole the treatment seems to work.

This seems very confusing, and just think how many medical reports in newspapers contain results of this type: drinking red wine is good for you, eating meat is bad for you, and so on. What has gone wrong?

The key to this paradox is to note that the majority of the older patients are actually in the treatment group. This confuses the effect of the treatment with a perfectly possible dependence on the age of the recipient. In essence this is a badly designed trial, but there is no doubting that it is a subtle effect and not one that most people could understand without a great deal of careful explanation.

The Curious Case of the Inexpert Witness

All this brings me to the tragedy that was largely responsible for me deciding to write this book. In 1999, Mrs Sally Clark was tried and convicted for the murder of her two sons Christopher, who died aged 10 weeks in 1996, and Harry who was only eight weeks old when he died in 1998. Sudden infant deaths are sadly not as uncommon as one might have hoped: about one in eight thousand families experience such a nightmare. But what was unusual in this case was that after the second death in Mrs Clark's family, the distinguished paediatrician Sir Roy Meadow was asked by the police to investigate the circumstances surrounding both her losses. Based on his report, Sally Clark was put on trial for murder. Sir Roy was called as an expert witness. Largely because of his testimony, Mrs Clark was convicted and sentenced to prison. After much campaigning, she was released by the Court of Appeal in 2003. She was innocent all along. On top of the loss of her sons, the courts had deprived her of her liberty for four years. The whole episode was a disgrace.

I am not going to imply that Sir Roy Meadow bears sole responsibility for this fiasco, because there were many difficulties in Mrs Clark's trial. One of the main issues raised on Appeal was that the pathologist working with the prosecution had failed to disclose evidence that Harry was suffering from an infection at the time he died. Nevertheless, what Professor Meadow said on oath was so shockingly stupid that he fully deserves the vilification with which he was greeted after the trial. Two other women had also been imprisoned in similar circumstances, as a result of his intervention.

At the core of the prosecution's case was a probabilistic argument that would have been torn to shreds had any competent statistician been called to the witness box. Sadly, the defence counsel seemed to believe it as much as the jury did, and it was never rebutted. Sir Roy stated, correctly, that the odds of a baby dying of sudden infant death syndrome (or 'cot death') in a family were 8543 to one against. He then presented the probability of this happening twice in a family as being this number squared, or 73 million to one against. In the minds of the jury this became the odds against Mrs Clark being innocent of a crime.

That this argument was not effectively challenged at the trial is staggering. Remember that the product rule for combining probabilities $P(A \cap B) = P(A)P(B \mid A)$ only reduces to $P(A)P(B)$ if the two events are independent. Nobody knows for sure what causes cot deaths, but there is every reason to believe that there might be inherited or environmental factors that might cause such deaths to be more frequent in some families than in others. In other words, sudden infant deaths might be correlated rather than independent. Furthermore, there is data about the frequency of multiple infant deaths in families. The conditional frequency of a second such event following an earlier one is not 1 in 8000 or so, it is just 1 in 77. This is hard evidence that should have been presented to the jury. It was not.

Defending himself, Professor Meadow tried to explain that he had not really understood the statistical argument he was presenting, but was merely repeating for the benefit of the court something he had read, which turned out to have been in a report that had not even been published at the time of the trial. He said 'To me it was like I was quoting from a radiologist's report or a piece of pathology. I was quoting the statistics, I was not pretending to be a statistician.' I always thought that expert witnesses were suppose to testify about those things that they were experts about, rather than subjecting the jury to second-hand flummery. Perhaps expert witnesses enjoy their status so much that they feel they cannot make mistakes about anything.

Subsequent to Mrs Clark's release, Sir Roy Meadow (who was aged 72 at the time of writing) was summoned to appear in front of a disciplinary tribunal at the General Medical Council. At the end of the hearing he was found guilty of serious professional misconduct, and struck off the medical register. Since he is retired anyway, this seems scant punishment. The judges and barristers who should have been alert to this miscarriage of justice have escaped censure altogether.

Although I am pleased that Professor Meadow has been disciplined in this fashion, I also hope that the General Medical Council does not think that hanging one individual out to dry will solve this problem. I addition, I think the politicians and legal system should look very hard at what went wrong in this case (and others of its type) to see how the probabilistic arguments that are essential in the days of forensic science can be properly incorporated in a rational system of justice. At the moment there is no agreed protocol for evaluating scientific evidence before it is presented to court. It is likely that such a body might have prevented the case of Mrs Clark from ever coming to trial. Scientists frequently seek the opinions of lawyers when they need to, but lawyers seem happy to handle scientific arguments themselves even when they do not understand them at all.

Science, Society, and Statistics

I often think of the law courts as a sort of microcosm of human society. Accordingly, many of the points I have made about probability in the witness box apply to many facets of everyday life, including business, commerce, transport, the media, and politics. They even play a role in personal relationships, though only probably at a subconscious level. It is a feature of everyday life that science and technology are deeply embedded in every aspect of what we do each day. Science has given us greater levels of comfort, better health care, and a plethora of labour-saving devices. It has also given us unprecedented ability to destroy the environment and each other, whether through accident or design.

Society faces rigorous challenges over the next century. We must confront the threat of climate change and forthcoming energy crises. We must find better ways of resolving conflicts peacefully lest nuclear or conventional weapons lead us to global catastrophe. We must stop large-scale pollution or systematic destruction of the biosphere that nurtures us. And we must do all of these things without abandoning the many positive things that science has brought us. Abandoning science and rationality by retreating into religious or political fundamentalism would be a catastrophe for humanity.

Unfortunately, recent decades have seen a wholesale breakdown of trust between scientists and the public at large. This is due partly to the deliberate abuse of science for immoral purposes, and partly to the sheer carelessness with which agencies exploited scientific discoveries without proper evaluation of the risks involved. But more fundamentally it is due to an increasing alienation between scientists and the general public. Each year fewer students enrol for courses in physics and chemistry. Fewer graduates mean fewer qualified science teachers in schools. This is a vicious cycle that threatens our future and it must be broken.

The danger is that the decreasing level of understanding of science in society means that knowledge (as well as its consequent power) becomes concentrated in the minds of a few individuals. This could have dire consequences for the future of our democracy. Even as things stand now, very few Members of Parliament are scientifically literate. How can we expect to control the application of science when the necessary understanding rests with an unelected 'priesthood' that is hardly understood by, or represented in, our democratic institutions?

Very few journalists or television producers know enough about science to report sensibly on the latest discoveries or controversies. As a result, important matters that the public needs to know about do not appear at all in the media, or if they do it is in such a garbled fashion that they do more harm than good. I have listened many times to radio interviews with scientists on the Today programme on BBC Radio 4. I even did such an interview once. It is a deeply frustrating experience. The scientist usually starts by explaining what the discovery is about in the way a scientist should, with careful statements of what is assumed, how the data is interpreted, and what other possible interpretations might be. The interviewer then loses patience and asks for a yes or no answer. The scientist tries to continue, but is badgered. Either the interview ends as a row, or the scientist ends up stating a grossly oversimplified version of the story.

Some scientists offer the oversimplified version at the outset, of course, and these are the ones that contribute to the image of scientists as priests. Such individuals often believe in their theories in exactly the same way that people believe in fundamentalist religion.

Not with the conditional and possibly temporary belief that characterizes the scientific method, but with the unquestioning fervour of an unthinking zealot. This approach may pay off for the individual in the short term, in popular esteem and media recognition—but when it goes wrong it is science as a whole that suffers. When a result that has been proclaimed certain is later shown to be false, the result is widespread disillusionment.

The worst example of this tendency that I can think of is the constant use of the phrase 'Mind of God' by theoretical physicists to describe fundamental theories. This is not only meaningless but also damaging. As scientists we should know better than to use it. Our theories do not represent absolute truths: they are just the best we can do with the available data and the limited powers of the human mind. We believe in our theories, but only to the extent that we need to accept working hypotheses in order to make progress. Our approach is pragmatic rather than idealistic. We should be humble and avoid making extravagant claims that can not be justified either theoretically or experimentally.

The more that people get used to the image of 'scientist as priest' the more dissatisfied they are with real science. Most of the questions asked of scientists simply can not be answered with 'yes' or 'no'. This leaves many with the impression that science is very vague and subjective. The public also tend to lose faith in science when it is unable to come up with quick answers. Science is a process, a way of looking at problems, not a list of ready-made answers to impossible problems. Of course it is sometimes vague, but I think it is vague in a rational way and that's what makes it worthwhile. It is also the reason why science has led to so many objectively measurable advances in our understanding of the World.

I do not have any easy answers to the question of how to cure this malaise, but do have a few suggestions. It would be easy for a scientist such as myself to blame everything on the media and the education system, but in fact I think the responsibility lies mainly with ourselves. We are usually so obsessed with our own research, and the need to publish specialist papers by the lorry-load in order to advance our own careers that we usually spend very little time explaining what we do to the public. I think every working scientist in the country should be required to spend at least 10% of his time working in schools or with the general media on 'outreach'. People in my field—astronomers and cosmologists—do this quite a lot, but these are areas where the public has some empathy with what we do. If only biologists, chemists, nuclear physicists and the rest were viewed in such a friendly light. Doing this sort of thing is not easy, especially when it comes to saying something on the radio that the interviewer does not want to hear. Media training for scientists has been a welcome recent innovation for some branches of science, but most of my colleagues have never had any help at all in this direction.

The second thing that must be done is to improve the dire state of science education in schools. Over the last two decades the national curriculum for British schools has been dumbed down to the point of absurdity. Pupils that leave school at 18 having taken 'Advanced Level' physics do so with no useful knowledge of physics at all, even if they have obtained the highest grade. I do not at all blame the students for this. It's all the fault of the educationalists, who have done the best they can for a long time to convince our young people that science is too hard for them. Science can be difficult, of course, and not everyone will be able to make a career out of it. But that does not mean that it should not be taught properly to those that can take it in. If some students find it is not for them, then so be it. I always wanted to be a musician, but never had the talent for it.

I realize I must sound very gloomy about this, but I do think there are good prospects that the gap between science and society may gradually be healed. The fact that the public distrust scientists leads many of them to question us, which is a very good thing. They should question us and we should be prepared to answer them. If they ask us why, we should be prepared to give reasons. If enough scientists engage in this process then what will emerge is an understanding of the enduring value of science. I do not just mean through the DVD players and computer games it has given us, but through its cultural impact. It is part of human nature to question our place in the Universe, so science is part of what we are. It gives us purpose. But it also shows us a way of living our lives. Except for a few individuals, the scientific community is tolerant, open, internationally-minded, and imbued with a philosophy of cooperation. It values reason and looks to the future rather than the past. Scientists like anyone else will always make mistakes, but we can always learn from them. The logic of science may not be infallible, but it's probably the best logic there is in a world so filled with uncertainty.

Index